MU120
Open Mathematics

Unit 6

Maps

MU120 course units were produced by the following team:

Gaynor Arrowsmith (Course Manager)
Mike Crampin (Author)
Margaret Crowe (Course Manager)
Fergus Daly (Academic Editor)
Judith Daniels (Reader)
Chris Dillon (Author)
Judy Ekins (Chair and Author)
John Fauvel (Academic Editor)
Barrie Galpin (Author and Academic Editor)
Alan Graham (Author and Academic Editor)
Linda Hodgkinson (Author)
Gillian Iossif (Author)
Joyce Johnson (Reader)
Eric Love (Academic Editor)
Kevin McConway (Author)
David Pimm (Author and Academic Editor)
Karen Rex (Author)

Other contributions to the text were made by a number of Open University staff and students and others acting as consultants, developmental testers, critical readers and writers of draft material. The course team are extremely grateful for their time and effort.

The course units were put into production by the following:

Course Materials Production Unit (Faculty of Mathematics and Computing)

Martin Brazier (Graphic Designer)
Hannah Brunt (Graphic Designer)
Alison Cadle (TEXOpS Manager)
Jenny Chalmers (Publishing Editor)
Sue Dobson (Graphic Artist)
Roger Lowry (Publishing Editor)
Diane Mole (Graphic Designer)
Kate Richenburg (Publishing Editor)
John A.Taylor (Graphic Artist)
Howie Twiner (Graphic Artist)
Nazlin Vohra (Graphic Designer)
Steve Rycroft (Publishing Editor)

This publication forms part of an Open University course. Details of this and other Open University courses can be obtained from the Student Registration and Enquiry Service, The Open University, PO Box 197, Milton Keynes MK7 6BJ, United Kingdom: tel. +44 (0)845 300 6090, email general-enquiries@open.ac.uk

Alternatively, you may visit the Open University website at http://www.open.ac.uk where you can learn more about the wide range of courses and packs offered at all levels by The Open University.

To purchase a selection of Open University course materials visit http://www.ouw.co.uk, or contact Open University Worldwide, Walton Hall, Milton Keynes MK7 6AA, United Kingdom, for a brochure: tel. +44 (0)1908 858793, fax +44 (0)1908 858787, email ouw-customer-services@open.ac.uk

The Open University, Walton Hall, Milton Keynes, MK7 6AA.

First published 1996. Second edition 2006. Third edition 2008.

Edited, designed and typeset by The Open University, using the Open University TEX System.

Printed and bound in the United Kingdom by The Charlesworth Group, Wakefield.

ISBN 978 0 7492 2864 4

3.1

Contents

Study guide

This area of the OS map may be accessed via the OS website (see Stop Press for details).

Your study of *Unit 6* includes a new component—an extract from an Ordnance Survey (OS) map. It is used in Sections 2, 3 and 4.

Section 1 considers the general use of maps. Some short reader articles are integrated into this first section.

If you are unaccustomed to map reading, Sections 2 and 3 may take a longer time to study than the other sections, as much of your work will involve activities using the OS map extract. You will need to lay the map on a flat surface in good lighting. Section 3 also includes a video band, 'Getting your bearings'. Ideally, you should watch this before you begin to study the section. You will need a protractor to complete some of the activities and there is an audio sequence for those not familiar with using this measuring device.

Section 4 introduces you to the technique of drawing profiles—graphs that represent terrain. So you will need centimetre graph paper. Using your calculator to draw profiles and other graphs is the focus of Section 5, which also consolidates your learning.

Unit 6 uses metric or S.I. units for the measurement of distance: metres m, kilometres km, centimetres cm and millimetres mm. If you are not completely happy with this system of units, you may find it helpful to refer back to the preparatory resource books.

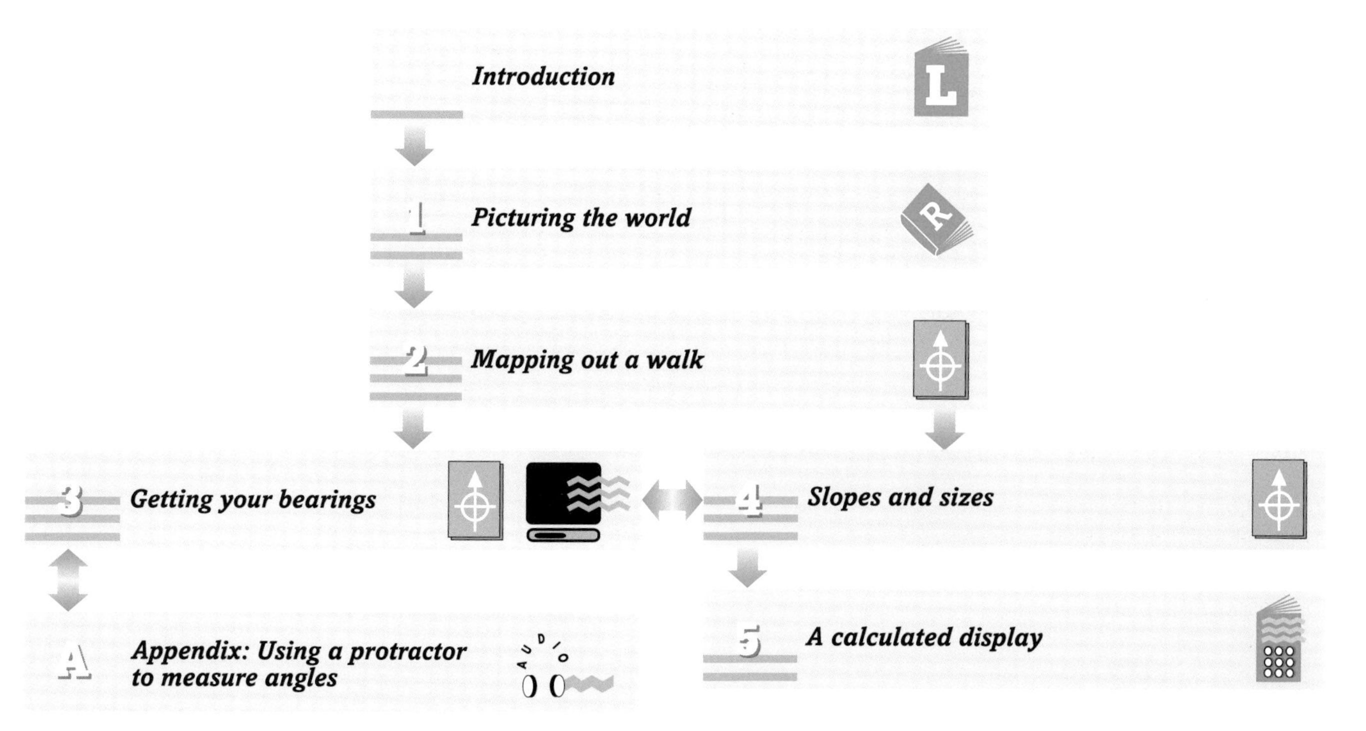

Summary of sections and other course components needed for *Unit 6*

Introduction to Block B

This block presents a new viewpoint in your mathematical studies. So far, you have been looking at the world from a statistical and largely numerical point of view—working with numbers such as means and medians to describe and characterize sets of data. But mathematics does not deal only with numbers: it uses a wide range of symbols to express and manipulate ideas and relationships. This block takes the next step in developing ways of seeing things mathematically using different methods of representing information. Many representations use pictures—graphic images, like maps or graphs—to represent the "story". Others use the language of numbers and symbols to express relationships.

There are four units in this block. *Unit 6* takes the context of maps to explore the symbolic language used to represent features of the landscape: a visual way of presenting information that may be difficult to put into words.

Unit 7 considers another form of symbolism—that displayed by graphical representation. In mathematics, graphs are used to give a visual impression of relationships, and are drawn according to certain conventions. You need to understand and work with these conventions to interpret a graph and to draw or display useful graphs of your own by hand and using your calculator.

Unit 8 develops algebra (using letters and other symbols) to describe general mathematical relationships. Looking at relationships algebraically is a way of moving from a specific to a more general case: a way of looking at how calculations will go in general rather than working out a specific numerical result. It provides a way of writing down relationships that would be quite complicated if expressed in words. But algebra is more than a convenient shorthand: it leads to ways of rethinking and re-organizing relationships. By following mathematical rules, symbolic expressions can be *manipulated* to reveal new relationships or to solve new problems.

Recall in *Unit 1* that generalizing relationships is one of the central features of mathematics.

Finally, *Unit 9* focuses on music as a source of mathematical ideas about representation and relationships. The unit looks at the mathematics embedded in the notions of musical scales, pitch and intervals. (Do not worry, you need no special musical knowledge to enjoy and learn from this unit.) This unit also consolidates your learning on the block.

Throughout this block you are asked to think about whether representations incorporate or reflect a bias or a particular viewpoint. A key theme is looking out for what is stressed and what is ignored in any representation.

Introduction

This unit concerns the representations and relationships related to maps. Maps use a particular language of special symbols to tell their stories. To read and interpret a map, you need to understand what the symbols represent.

Section 1 looks at a variety of maps and considers how they were designed and then used. Sections 2 and 3 examine one particular map: an Ordnance Survey (OS) map. Using this map involves a number of mathematical skills, as does Section 4, which considers how to quantify the steepness of a slope.

Before you begin, take a few moments to revise aspects of your work on Block A, which you will need here. You may find the activity sheet useful for this.

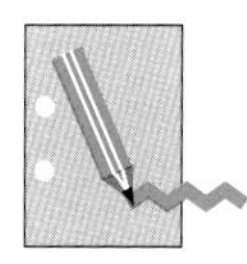

Activity 1 Reviewing topics from Block A

(a) The statistical and graphical features of the course calculator (e.g. inputting data into lists, analysing the data, representing it graphically) are important in Block B . Take some time to review your learning. Write notes on the following.

(i) Any relevant statistical, graphical or calculator skills you had before you started studying Block A.

(ii) The skills you have developed, during your study of Block A, in using the statistical and graphical features of the calculator. Give references to where you learnt these skills and evidence as to your current level of skill (e.g. activity numbers, TMA or CMA questions).

(iii) Describe at least one factor that helped you learn these skills, e.g. a way of working from the different course materials.

(b) The concept of ratio is also important in Block B.

(i) Look through your own work on Block A and select examples of your own work that illustrate your understanding of ratio. Choose at least one example and explain how ratio was used.

(ii) In your own words explain:

- what a ratio is and how it is calculated
- how it can be used.

You should refer to your example(s) in (b)(i) and at least two other uses of ratio from MU120, so far.

1 Picturing the world

Aims The main aim of this section is to introduce you to the idea that all maps are particular representations, and in using them you need to be aware of features that are stressed and those that are ignored. ◇

Much of this section provides background information which you are not expected to remember. Cartography—the science and art of map drawing—is one of the oldest human activities. Some of the earliest known maps are preserved on ancient clay tablets from Babylonian times. From very early in human history, it seems there was a need to represent the relative positions of known places. Maps are representations and every individual map embodies *both* a particular way of seeing and understanding, and a particular interpretation of the place it is depicting.

The next time you look at a world map, take a close look at the Greenwich meridian. It is an imaginary line, against which reference point most of the world sets its clocks. It runs north–south through Greenwich, London, the birthplace of modern navigation. The Greenwich meridian and the equator are the reference for an imaginary grid of lines overlaying the surface of the globe. *Meridians*, or lines of longitude, run from North to South pole; *parallels*, or lines of latitude, run around the globe parallel to the equator. Based on this grid, mathematical skills were developed to improve global navigation.

1.1 Map uses and conventions

Maps are human creations. They express particular interpretations of the world, and they affect how people understand that world. They are designed to serve a particular purpose. No map can show everything. Some things are selected and others omitted.

For example, the Ordnance Survey map extract you have been sent excludes information on the rainfall and details about the population of the area it covers. Some other maps would show such information.

It is the intended purpose of the map that determines what is to be included. For example, a simple map in a railway passenger timetable would not include lines used solely for freight.

Different people use maps in different ways and for different purposes. Sometimes these purposes diverge from those for which the map was created. An old map, created for navigational purposes, may today be hung on a wall as a work of art.

Activity 2 What is a map for?

Look at each of the maps in Figure 1. For each one, ask yourself for what purpose the map was created and how you might use that map.

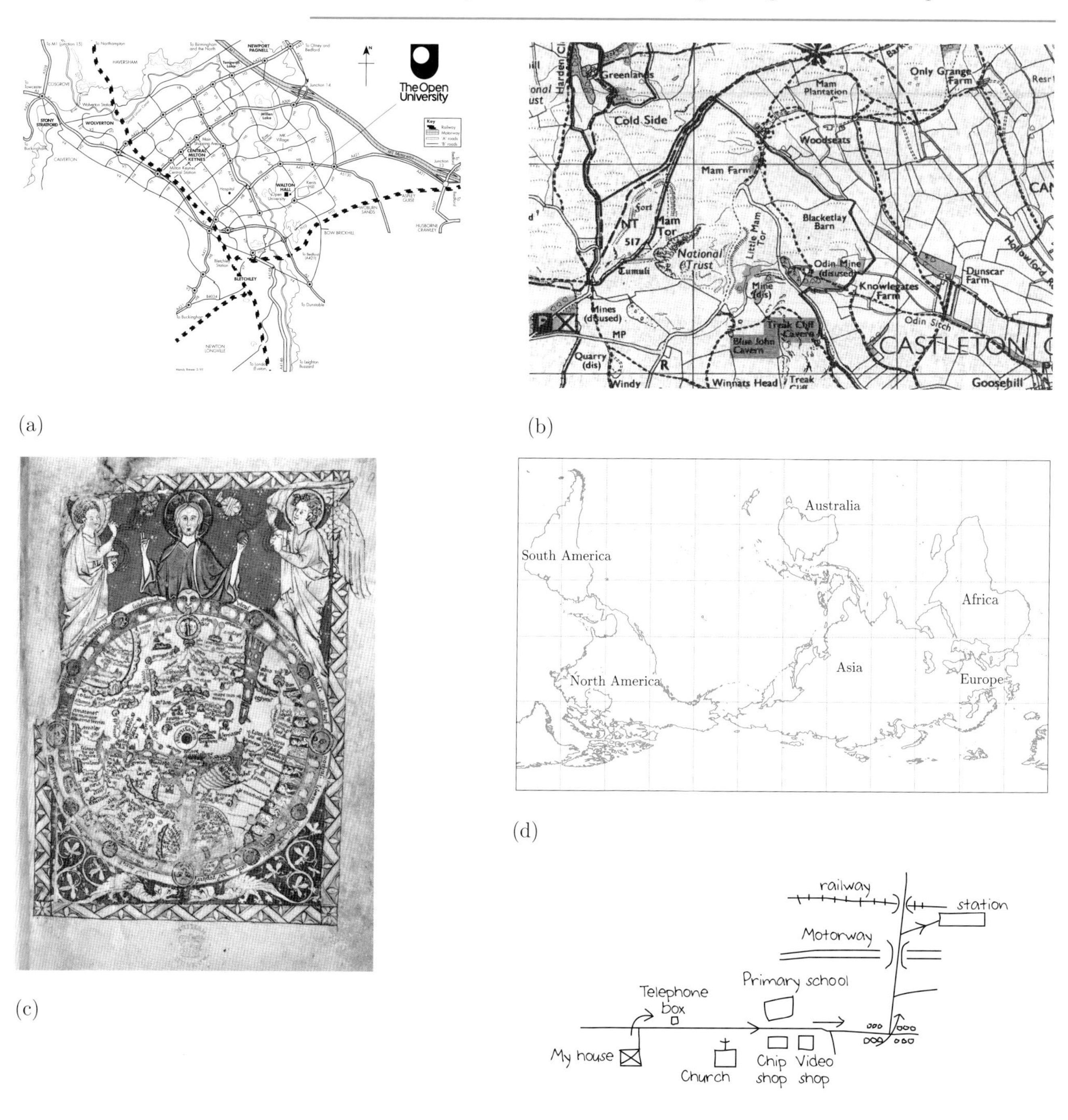

(e)

Figure 1 Maps for use with Activity 2

So what makes a map useful?

Usually a map needs to portray spatial information efficiently on a flat surface for users who may be engaged in a wide variety of activities—from recreation, education, legislation and decision-making, to navigation. To achieve this goal, particular conventions are used.

Conventions—styles of presentation—may include a title, orientation information (e.g. a north arrow), a scale, and a key for any symbols used. You might like to look at the maps shown in Figure 1 to see if these conventions have been used. First, consider the orientation of a map.

The orientation of a map

When people open up a map, they often take for granted that north is at the top of the page and south is at the bottom—it is orientated north–south. Today most maps, like Figure 1(a) and (b), obey this convention. But have you ever wondered why?

There is no 'right' way up for a map. You may find that, when you are using a map to follow a route, you actually turn the map round so that its orientation is the same as your own.

The very word 'orientation' reveals that some of the earliest maps pointed east (*oriens* in Latin) as does the Psalter world map of about 1260, shown in Figure 1(c). In Figure 1(d), south is at the top.

The north orientation of most maps is no more than a convention, which may have arisen from the development of compasses and their behaviour relating to the Earth's magnetism.

> ### *Historical note*
>
> Around 2500 years ago, the Greeks discovered the magnetic properties of lodestone—the mineral now called *magnetite*, which is an iron ore. By the first century AD the Chinese had used it to construct the first compass in the form of a lodestone spoon balanced on a smooth plate. The spoon moved to point north–south. Travellers of the fourteenth century brought this discovery back to Europe, where the device was adopted as a navigational aid and played a major part in the voyages of such explorers as Columbus and Magellan.

Today, many people are unsettled by anything other than a north-orientated map—demonstrating the strengths of present convention! However, other orientations are still used. The 1979 Australian map of the world entitled 'McArthur's universal corrective map of the world' (Figure 1(d)) has south at the top of the page. This map reflects the increasing importance that Australia gives to being in the southern hemisphere, rather than looking so much towards Britain and its colonial past in the northern hemisphere.

Conventions for directions

'Cardinal' here means 'of primary importance'.

A traditional way of specifying direction used in the past for navigation at sea was the four cardinal points of the compass: north, south, east and west. These were subdivided to give the directions north-east, south-east, south-west and north-west, as in Figure 2.

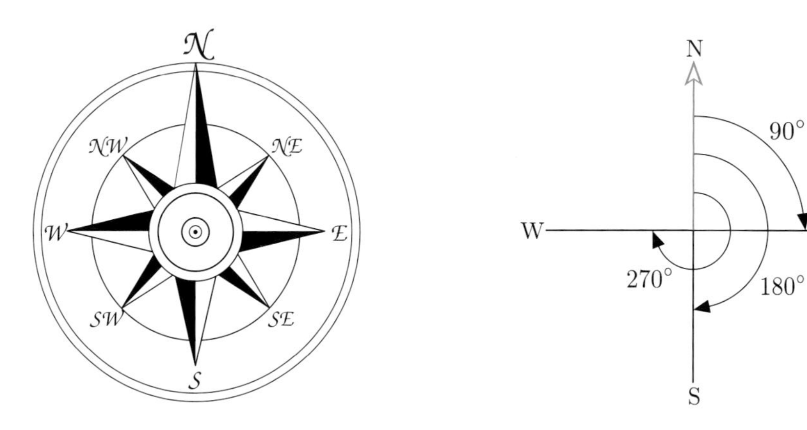

Figure 2 Named points of the compass

Figure 3 Compass directions using angles

Further subdivisions give further named points of the compass. For example, between north and north-east lie the directions north-by-east, north-north-east and north-east-by-north. Accurately specifying a direction quickly becomes quite complicated! However, there is a different numerical approach:

> the angle measured clockwise in degrees from north to the relevant direction

as shown in Figure 3. A direction of 90 degrees corresponds to east; 180 degrees is south; and 270 degrees is west. A full circle is 360 degrees. So a bearing of 360 degrees is the same as 0 degrees, and corresponds to north.

Activity 3 Compass bearings

What are the bearings in degrees that correspond to the four compass directions north-east, south-east, south-west and north-west?

Measuring angles clockwise from north is a convention among map-users. It is, however, not the same as the convention used in geometry, where angles are measured anticlockwise. Different groups of people do things in different ways—just as the customs of one country may differ from those of another.

1.2 *Squaring the circle*

The graphical and mathematical problems of representing the (roughly spherical) surface of the Earth as a flat map have taxed the ingenuity of map-makers from ancient times. All maps are compromises, capable of portraying one or more of the Earth's features, yet incapable of showing accurately on the same flat surface all four essential elements: true *shape*; equal *area*; accurate *distance*; and consistent *orientation*. Only globes are capable of reflecting all these features, but globes have disadvantages compared to flat maps. Even the largest globes have a very small scale and therefore show relatively little detail. Furthermore, globes are costly to make, difficult to store, heavy and impractical to carry around.

Current mappings of the world draw both on the accumulated knowledge of centuries and on modern high-technological processes. And yet there are still debates about how the world can best be represented. There is a whole area of mathematical study related to map projections—how to resolve the difficulty of representing a curved or spherical surface on a flat surface. Some kind of choice always has to be made. Different map projections reinforce different choices, and so different views of the world.

The next time you peel an orange, cut the peel off in one piece and arrange it on a flat surface. Think about the different ways you could do this—one way is shown in Figure 4.

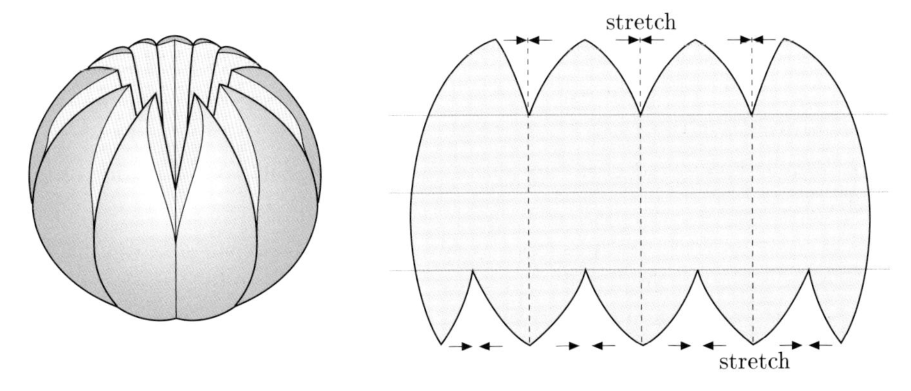

Figure 4 Orange peel arranged on a flat surface—a projection

This is a particular projection of a broadly spherical object onto a flat surface. Just as there are different ways to peel an orange, so there are different map projections of the Earth. Most of these involve a 'cylindrical' map projection. One such projection is shown in Figure 5 overleaf.

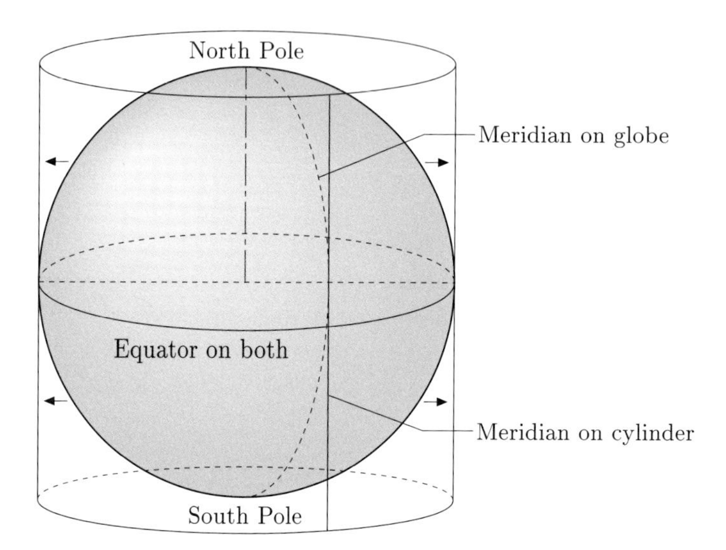

Figure 5 A cylindrical projection

A cylinder is wrapped round the globe. The lines of latitude and longitude are transferred onto the cylinder, which can then be opened out flat to form a map sheet. There are different ways of projecting a sphere onto a cylinder.

During the sixteenth century Gerard Mercator (1512–1594), a Flemish map-maker, mathematician and instrument maker, developed his world projection to facilitate navigation. His projection drew the meridians as equally spaced straight lines at right angles to the parallels, creating a grid called a *graticule* (Figure 6). This construction allowed any constant compass bearing to be plotted as a straight line on the map.

However, preserving true compass directions meant sacrificing true distance, area and proportions of the Earth's landmasses. Note a massive enlarging towards the poles. Greenland, which would fit into South America over eight times, appears the larger of the two. The actual poles cannot be projected at all.

Aware of these limitations, other map-makers have employed different projections of the world's surface. The German Arno Peters (1916–2002) stressed the principle of 'equal area', representing equal areas on the ground by equal areas on the map. This projection gives the reader the correct impression of relative land masses, but distorts angles, directions and the shape of coastal outlines (Figure 7).

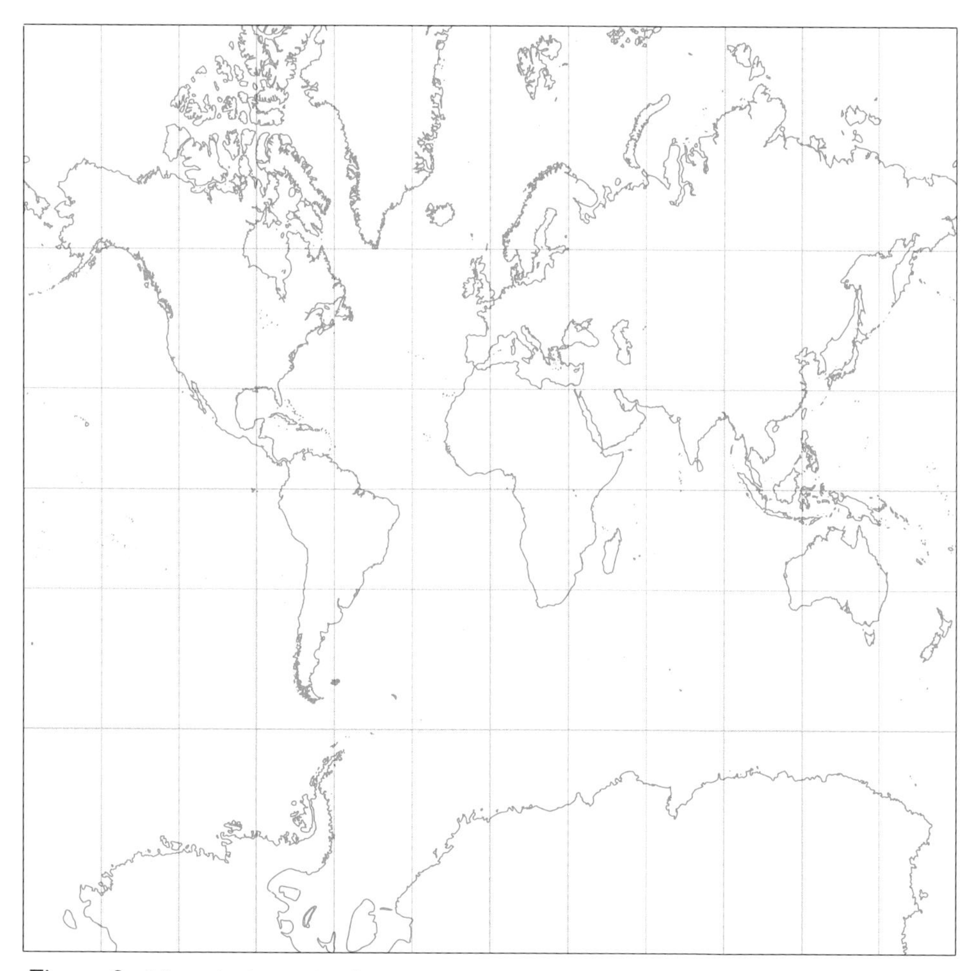

Figure 6 Mercator's projection

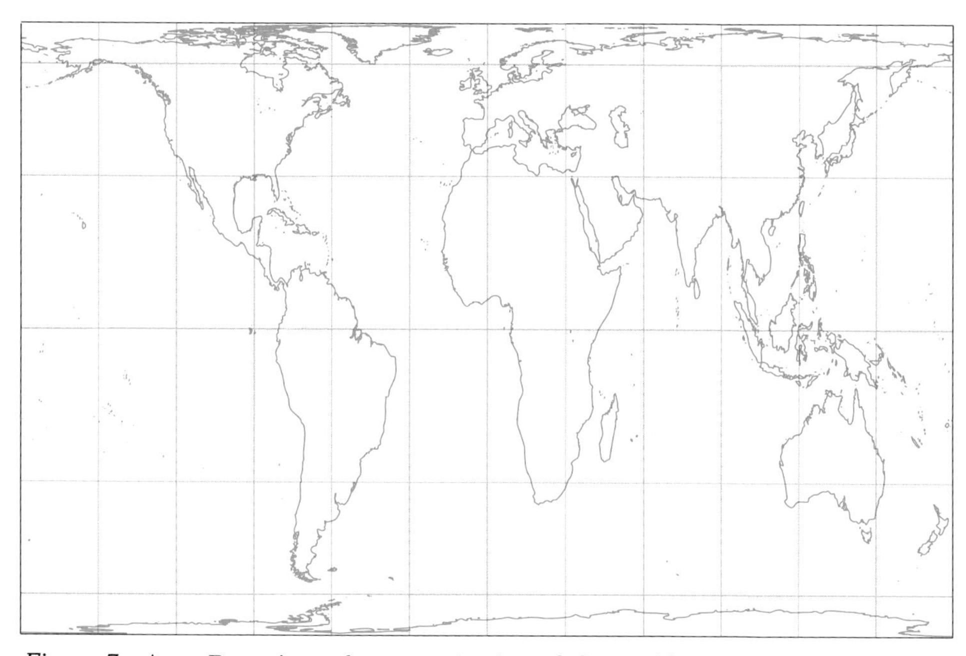

Figure 7 Arno Peters' equal-area projection of the world

In 1963, Arthur Robinson (1915–2004) preserved area in a different projection of the world (Figure 8), in which straight lines become curves.

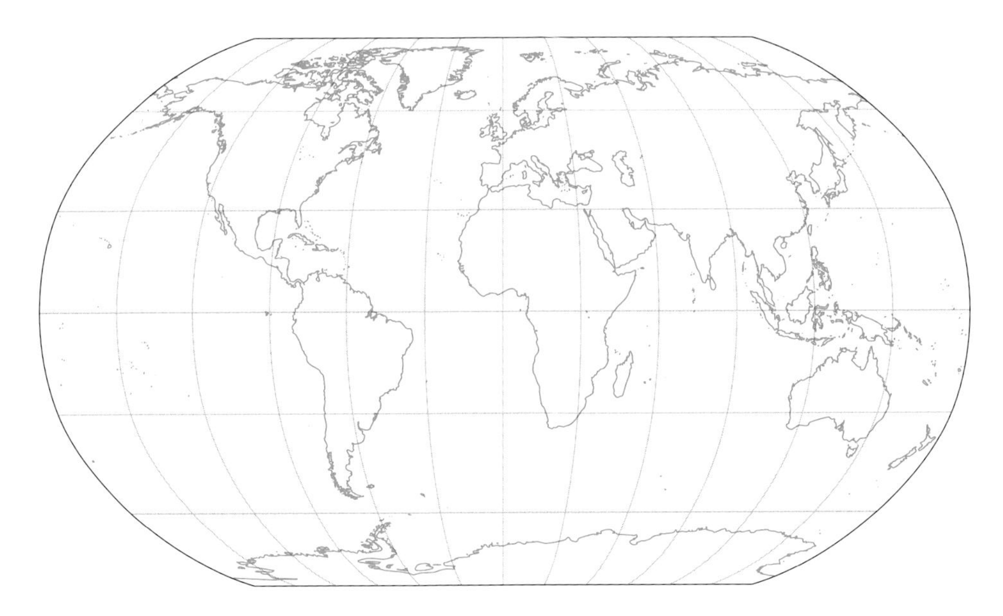

Figure 8 Robinson's projection

Other projections have tried to compromise between preserving true direction, distance, area and shape, by more complicated mathematical projections. For example, in the Hobo–Dyer projection, 'the cylinder' cuts through the globe at the 37.5° parallels but in order to preserve the equal area property the shapes of the land masses become progressively flattened towards the poles; the shapes of the land masses between 45° north and south are well preserved, however.

See http://www.odt.org/hdp/

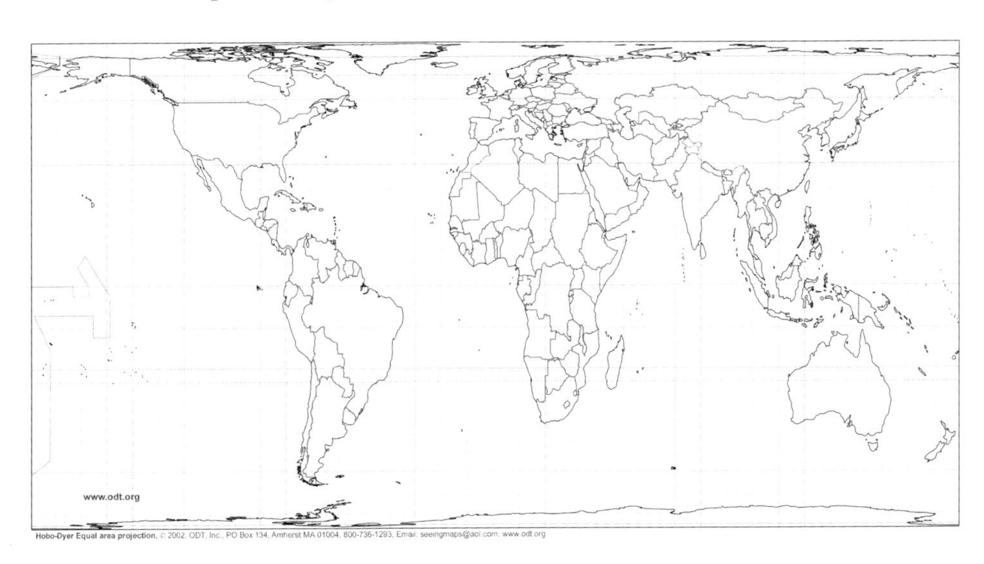

Figure 9 The Hobo–Dyer projection

To give a more local example, in 2005 the BBC were persuaded to amend the projection of their new UK weather maps, after many complaints by Scots that the area of their country looked disproportionately small compared with the south of England.

Activity 4 *Transforming 3D to 2D*

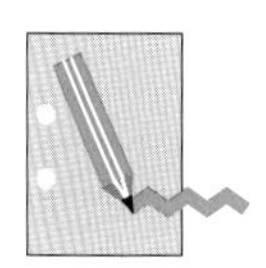

For your Handbook, make notes on the main elements that might be preserved when designing a representation of the 3D (three-dimensional) world on a flat piece of 2D (two-dimensional) paper.

What other examples can you think of where a 3D object is transformed to a 2D image?

1.3 Maps for particular purposes

For maps to be useful, it is important to be aware of what has been stressed, what has been ignored and the resulting limitations.

Activity 5 *Making choices and justifying decisions*

Imagine you are involved in making each of the following maps. State which principles and/or conventions you would consider most important to implement, giving your reasons.

(a) A sketch map for a friend to find her way from the nearest bus stop, when she comes to visit you.

(b) A map for an airline pilot to navigate on a flight over the North Pole.

(c) An Ordnance Survey map of a mountainous area.

(d) A world map showing the historical journeys of European explorers to Australia.

As you work through the next series of case studies, try to identify the particular problems faced by the map-makers and think about the way they were overcome, identifying the mathematical techniques used. You may find it useful to use the Handbook sheets and continue to add to your sheets as you work through the unit.

Case study 1 Network maps

Network maps demonstrate how a map may be a very simple representation used for a specific purpose. The information presented is highly selected so much is completely ignored.

Look at the map of the Singapore underground railway system (the Mass Rapid Transport or MRT) in Figure 10. Many rail networks produce such maps. The directions of lines and their twists and turns have mostly been removed. The purpose of the map is to help people negotiate the MRT system as quickly and easily as possible. The only important features are the stations and the connections between stations.

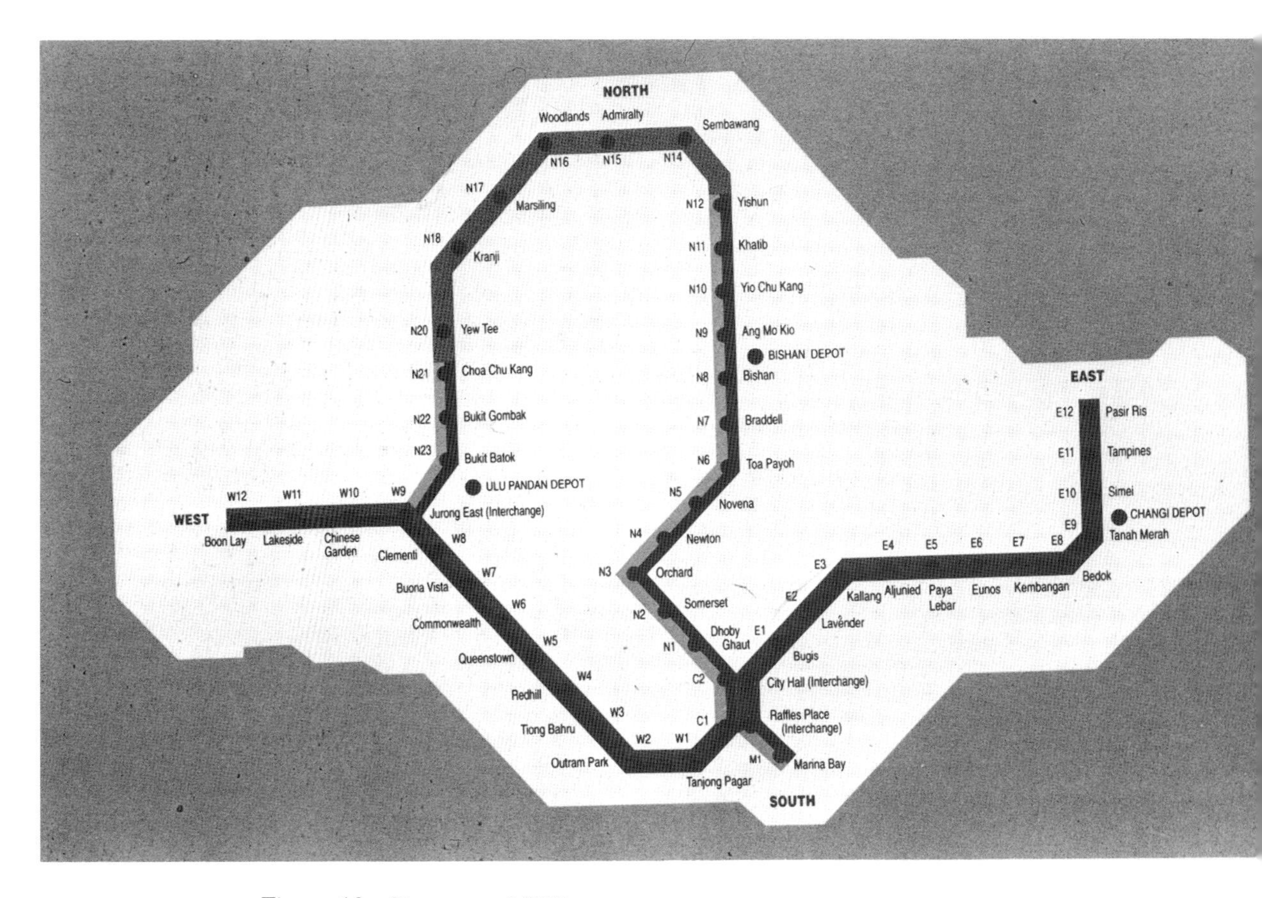

Figure 10 Singapore MRT map

There is distortion of relative directions in order to present a simple, easy to use map. Here there is no need to indicate the distance or time between stations as they are only a few minutes apart. Some road network maps do, however, show distances between junctions, as do some railway network maps. Others show the fastest times between major stations.

Network diagrams are often designed to aid route planning. Distribution agencies need to be able to identify and monitor particular routes. A conventional road atlas may present too much detail for this purpose. Network maps may be more appropriate.

Figure 11 shows how a network map can be created from a conventional road map. Detailed features such as bends in the road, bridges or other landmarks are omitted.

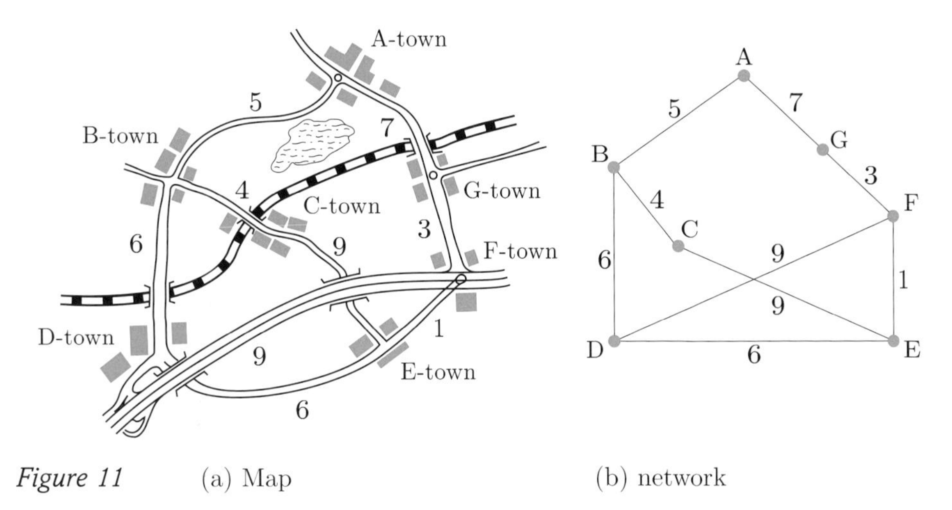

Figure 11 (a) Map (b) network

A network is a simple mathematical model with only the connectivity information preserved. It consists of points, called *vertices*, and lines, called *arcs*, which connect them. If there are numbers on the arcs (for example, to represent the distance or time between vertices), then these are called *weights*. (You may remember the idea of using numerical weights from *Unit 2*.) The terminology is common to a number of applications.

A mathematical model simplifies the real world.

Example 1 Calculating using a network diagram

Several towns, $A, B, \ldots$, are represented in the network in Figure 11(b). The 'weights' of the arcs are the distances, measured in kilometres, between towns. Notice that the network is not drawn to scale.

What distance do you travel if you travel to B from E via D?

B to D is 6 km and D to E is 6 km, so the total distance via this route is 12 km.

Activity 6 How far is it?

Use Figure 11(b) to find how far it is from B to E:

(a) via C

(b) via A, G and F.

Network maps may help to chart progress and aid planning, but in producing them a great deal of information is thrown away, while the essential spatial detail, represented using vertices and arcs, is retained.

► Can you identify what is stressed and what is ignored?

Networks stress the 'connectivity', i.e. whether objects are or are not connected. They may give distances or journey times, but they ignore all other features.

Case study 2 Tactile maps

The term 'tactile mapping' refers to any map in which touch is the predominant interpretative sense, such as maps that are specially designed for blind and partially-sighted people.

A *tactile map* has raised lines and symbols, so that you can *feel* rather than see the map. It is a very simple but useful idea. Maps for blind and partially-sighted people have been produced to provide them with freedom to travel or explore and be informed about new areas: for example, the layout of a railway station, airport terminal building, museum gallery or the route for a walk. Figure 12 shows a picture of a tactile map of a canal walk together with a sketch map of the same locality.

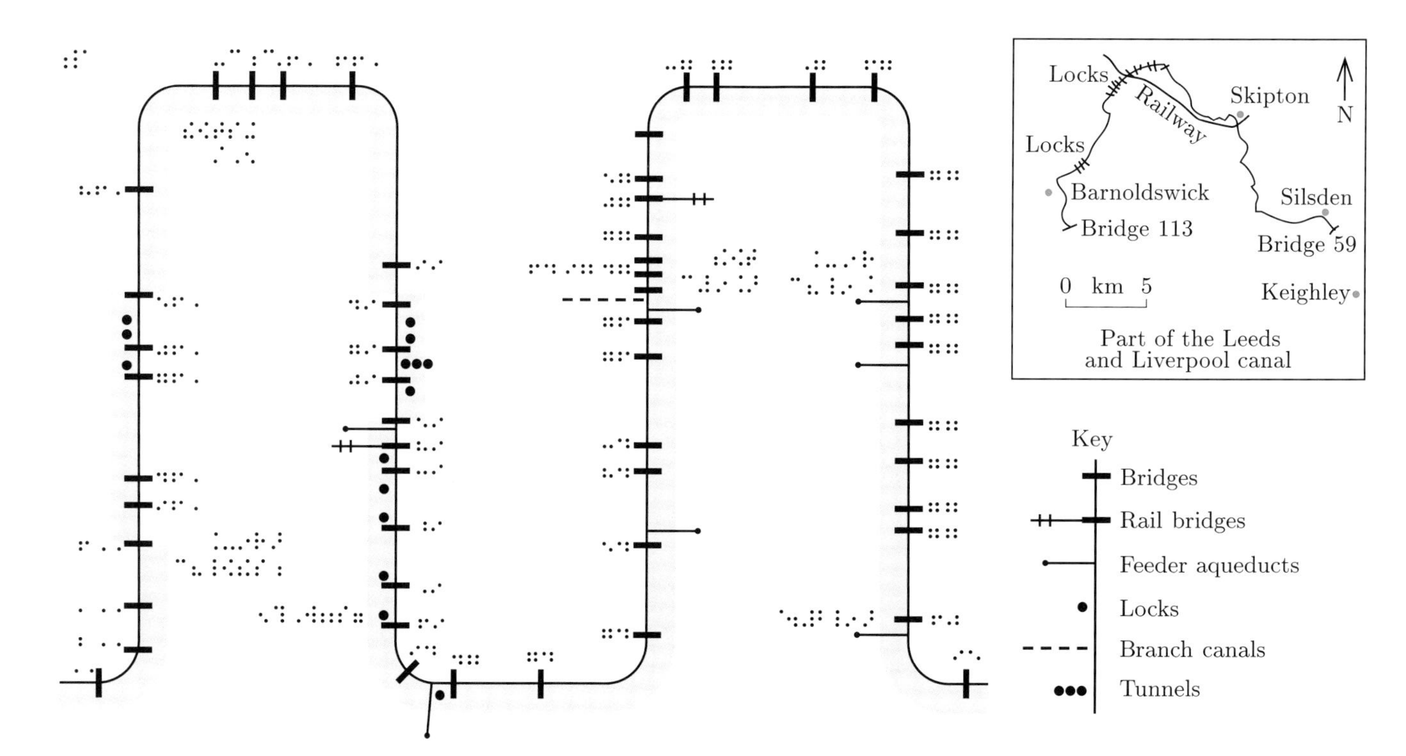

Figure 12 Diagram of a tactile map and sketch map

Activity 7 Principles of a tactile map

Study the map in Figure 12 and the associated sketch of the region on the right, then write down your observations on the principles involved in producing a tactile map. What features would you include in constructing a tactile map considering the needs of the intended user?

Case study 3 Mapping diseases

In *Unit 4*, you studied some maps created to show the spread of disease. John Snow's 1855 essay 'On the mode and communication of Cholera' included a map of Soho (London) on which he recorded cases of cholera. The map revealed a high concentration of the disease in the region of the Broad Street water pump, which supported his theory that cholera was passed on through drinking water. The water pump was chained up, which helped control the epidemic.

The mapping of the spread of disease can enable forward planning in terms of treatment, care, prevention and education.

Activity 8 Patterns of disease

Study the reader article 'Mapping the AIDS pandemic' by Peter Gould. What information has been collected? Can you suggest reasons for a reluctance to map the spread of AIDS in too much detail? What problem does this create for the representation of British cases?

Case study 4 Topographical maps

The geographical features of a land surface are called its *topography*. Maps showing these details are called *topographic maps*. Hills and valleys, undulations of the land surface, are known as the *relief*. One of the problems that early map-makers faced was how to portray the relief of a land surface on a flat piece of paper. This is a problem that has been tackled by different cultures in a variety of ways over the years.

> everyone agrees that *hills* are the hardest thing to get right on a map.
>
> (Denis Wood (1993) *The Power of Maps*, Routledge, London, p. 144)

On early maps, hills were shown in a very diagrammatic way as little 'cones' dotted about the map to represent where the top of the high ground lay. Later, map-makers used the technique of *hachuring*, in which the side of the hill is shaded to show slopes.

When the Ordnance Survey mapped Great Britain, the heights of prominent hills were measured relative to sea-level and marked on the maps.

The Ordnance Survey

In 1791, the Board of Ordnance was founded under the control of the British Army to provide more accurate and systematic maps for defence in the state of war between Britain and France.

The primary responsibility of the Ordnance Survey is to survey and provide maps of Great Britain. In early Victorian England, their maps came to be used for the management and transfer of land, civil engineering and efforts to improve the sanitary conditions of the growing industrial towns. Scientific uses, including geological and archaeological mapping, also developed, so that the Ordnance Survey provides a wide range of maps of Great Britain for use by the public and the government.

(Based on Harley, J. B. (1975) *Ordnance Survey Maps—a descriptive manual,* Ordnance Survey, Southampton, p. 3)

The word 'ordnance' refers specifically to military weapons, and more generally to military stores.

Until the mid-1970s, the post of Director-General of the Ordnance Survey was held by a senior army officer.

You will find spot heights on your OS map, but they do not carry the information about the shape of the countryside. That task is accomplished by the brown lines called *contour lines.* The most important step of all in topographic mapping was the representation of relief on maps by contours drawn at regular height intervals. Each contour on the map joins points which are a particular height above mean sea-level. Mean sea-level is used as the reference plane, and is known as the Ordnance Datum (OD).

The Ordnance Datum (OD) is the mean level of the sea at Newlyn in Cornwall. 'Datum' is the singular of 'data', a single item of numerical data.

The mathematical idea behind contours is not unique to the representation of height on the land surface. Lines joining places of equal numerical value with respect to a given climatic or other variable are called *isograms.* Examples in common use on weather maps include *isobars*, which join places of the same atmospheric pressure; *isotherms*, which are lines of equal temperature; and *isohyets*, which link places of equal precipitation (contour lines of equal rainfall).

The Greek prefix 'iso' means 'equal'.

Figure 13 shows a chronology of some of the ways that hills have been depicted graphically from the signs used by the Mixtec and Nahuatl peoples of pre-Hispanic Mexico some 6000 to 7000 years ago, to the more familiar representations in use today.

Case study 5 Digital maps

Geographical data are now available from 'Geographical Information Systems', or GIS, described in a HMSO publication as 'A system for capturing, storing, checking, integrating, manipulating, analysing and displaying data which are spatially referenced to the Earth'. GIS works by integrating all kinds of information into a single system, where everything can be referenced in terms of data on geographical position.

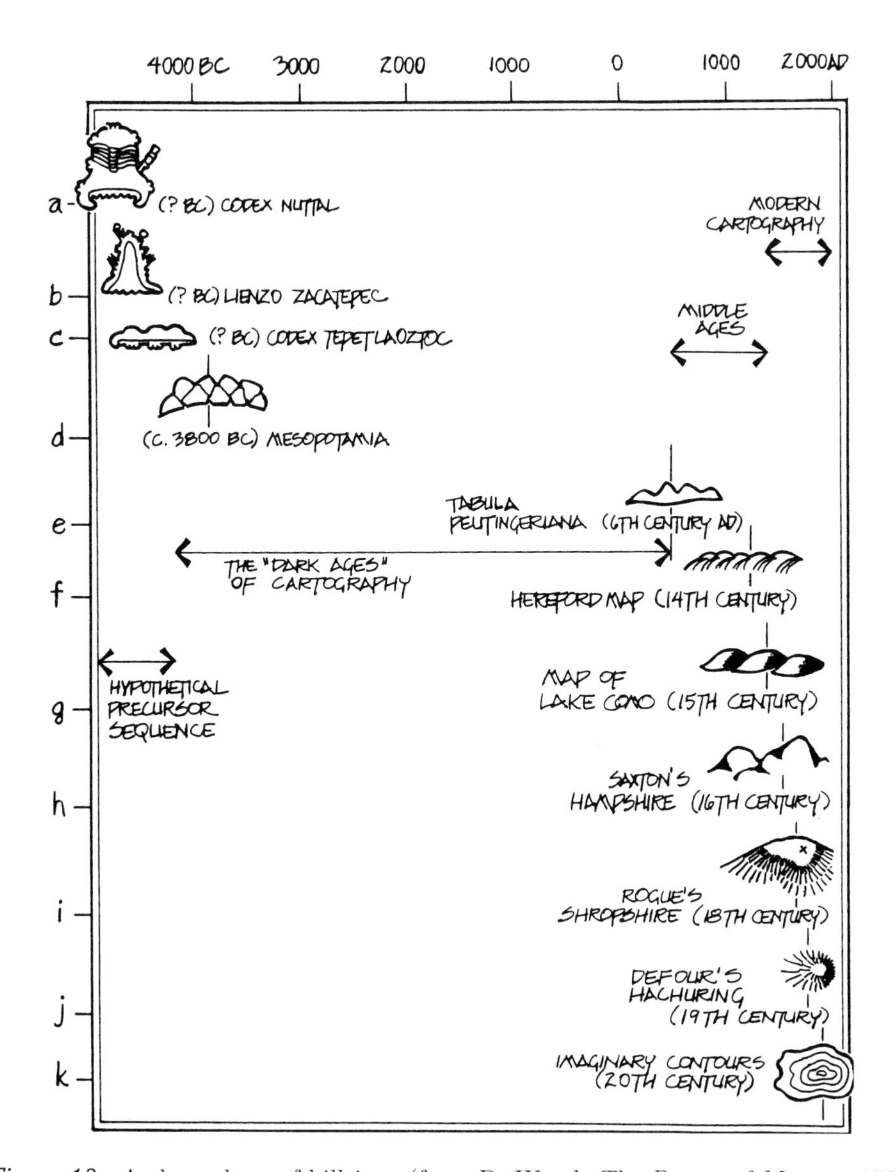

Figure 13 A chronology of hillsigns (from D. Wood, *The Power of Maps*, p. 153)

The data, in digital form, can be displayed in a variety of ways: as maps, as photographs, in tables, by postcode. It is a very powerful system. It can be used for many purposes from traffic planning to land-use disputes, to geological exploration. The Ordnance Survey now store all their maps digitally, and the maps can be accessed via the OS website. However, inevitably, not everything is included in any digital map. What is included reflects the interests of those who design, own and operate the system.

The Global Positioning System, or GPS, is often used in conjunction with digital maps for navigation systems. The GPS system tells the navigation system where you are, and the system uses its digital map to suggest a route to your destination. Data on traffic jams can be fed into such road navigation "maps" to make them even more useful.

Activity 9 Views about maps

To finish this section, study the reader article 'Images of the World' by Denis Wood. Do you agree with his views? Why does he comment that all maps are biased?

To summarize, maps are created for many different purposes, including navigation (finding your way), communicating information (such as climate information), aiding research into the spread of diseases, or making political points.

Different conventions are used in maps: for instance, nowadays north usually points to the top of the map. Conventions depend upon the purpose for which the map was created and upon those who created it.

The features that are included in a map depend upon the map's purpose, as do those features that are excluded. For instance, the features needed in a tactile map of an area for blind or partially-sighted people are different from those needed for sighted people.

One of the simplest forms of maps is a network. Networks show points (vertices) and links between the points (arcs).

In a 2D map of the 3D Earth, it is not possible to represent accurately shape, area, distance and orientation.

Each map necessarily has a particular perspective, stressing some things and ignoring others. Mathematics is an important and useful tool in map production. Data collection, designing maps for a specific purpose and creating an accurate image, all use mathematical ideas and skills.

Now check that you have completed your Handbook notes for this section.

Outcomes

After studying this section, you should be able to:

- ◇ review your learning so far (Activity 1);
- ◇ consider how the purpose of a map relates to the information it contains (Activities 2, 5, 7, 8 and 9);
- ◇ understand map conventions and specify direction using compass points and angles between 0° and 360° (Activities 3 and 5);
- ◇ describe the properties that cannot all be preserved in a transformation from a 3D to a 2D representation (Activities 4, 5 and 9);
- ◇ identify features in a representation that have been stressed and some of those that have been ignored (Activities 4, 5, 7 and 8);
- ◇ read maps, networks and text for a purpose and extract appropriate information (Activities 2, 6, 8 and 9).

2 Mapping out a walk

Aims The main aim of this section is to bring out mathematical ideas that are embedded in maps, such as the grid reference system, use of scales and contour line representations. ◇

In this section and the next you are going to use an extract from the Ordnance Survey (OS) Outdoor Leisure Map 1—the Dark Peak area map extract—to plan a walk in the Peak District National Park in the UK. If you visit this beautiful part of the country, you can try the walk yourself.

However, this section is not just about map reading. It is also about the mathematical features of maps. The skills you will be developing are part of a mathematical language that uses graphical representations to communicate information.

No prior knowledge of OS maps is assumed. If the ideas are all new to you, however, then your study time may be longer than for someone who is familiar with OS maps. The next activity is designed to help you assess your experience and so adapt your study plans accordingly.

Activity 10 Familiar maps?

Have you used an Ordnance Survey map before?

Using the criteria 'very familiar', 'familiar' or 'not at all familiar', how would you rate your skill in the following activities:

◇ using grid references;

◇ using the map scale to work out distances;

◇ interpreting contours?

2.1 Pointing by numbers

Look at the map mailed with this unit and keep it to hand as you work through this section.

The walk you will be planning in this section is in the hills north of Castleton.

▶ Can you find Castleton on your map or do you need more information?

▶ How can particular locations on a map be specified accurately and unambiguously?

With the map in front of you, you can point to a map feature and say 'That's where we're going'. But you cannot do this over the phone or in a letter. How can you convey the same information in these cases?

Note that the National Grid lines do not correspond to lines of either latitude or longitude.

Maps produced by the Ordnance Survey use a reference system called the National Grid (not to be confused with the UK's electrical power distribution system). It covers Great Britain with a grid of north–south and east–west lines. Figure 14(c) shows the grid subdividing the country into squares with 100 kilometre sides. The Grid forms a reference system, which can be used to locate any point in Great Britain.

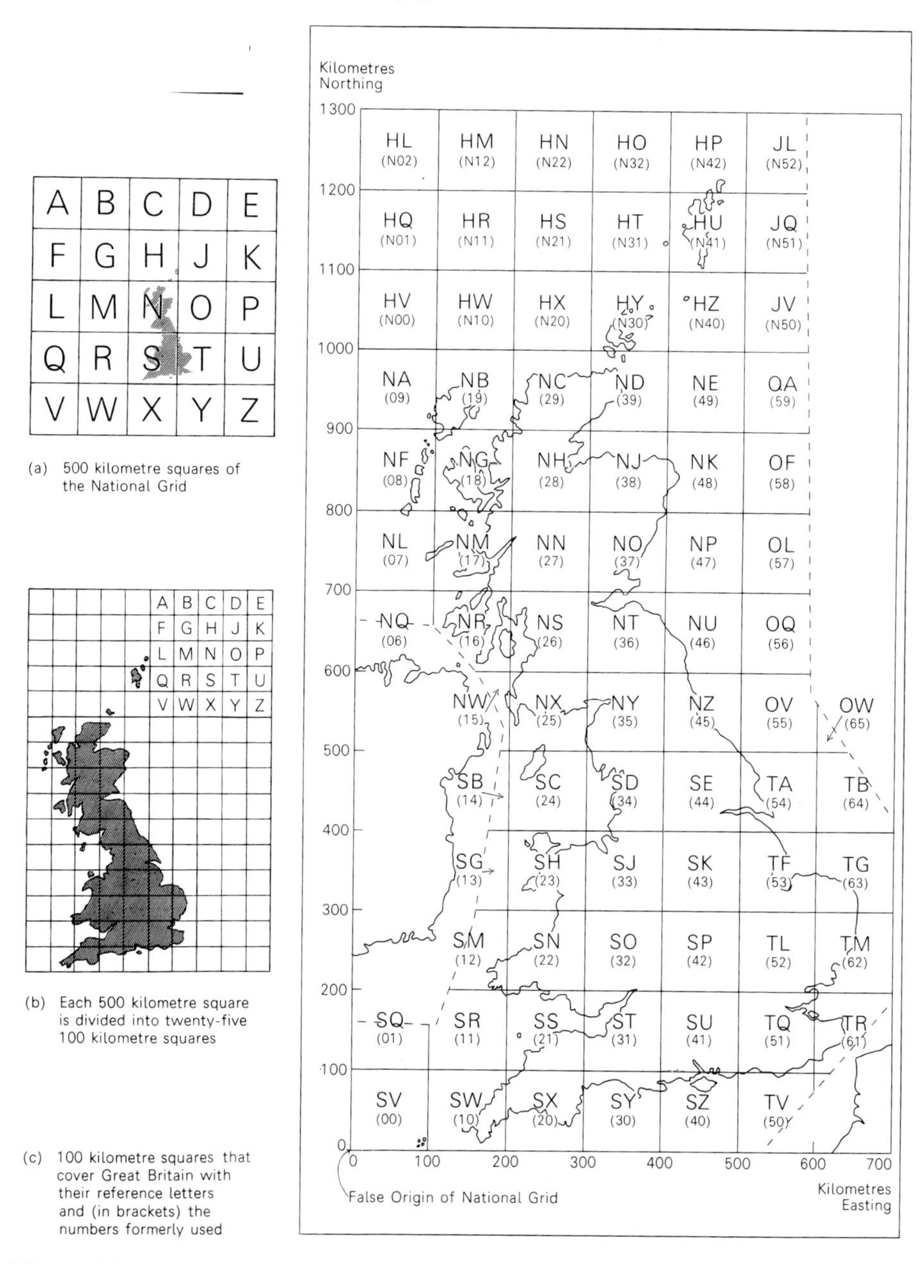

(a) 500 kilometre squares of the National Grid

(b) Each 500 kilometre square is divided into twenty-five 100 kilometre squares

(c) 100 kilometre squares that cover Great Britain with their reference letters and (in brackets) the numbers formerly used

Figure 14 National Grid reference system of Great Britain

Along the bottom and up the left-hand edge of Figure 14(c) you will see numbers. The numbers along the bottom show distances to the east, and are called *eastings*. The numbers up the left-hand edge show distances to the north and are called *northings*. The *origin* of the Grid, where both the easting and northing are zero, is at the bottom left-hand corner of Figure 14(c) about 100 kilometres to the west of the Scilly Isles.

Because it is part of a larger grid system, shown in Figure 14(a), the origin, shown in Figure 14(c), is referred to by the Ordnance Survey as a 'false origin'.

Each grid square is given a unique two-letter reference. The Peak District occupies part of two grid squares SK and SE on Figure 14(c). However, you can also specify each square using eastings and northings.

The convention is to take the easting first. Locate the north–south line that runs along the left-hand edge of the square SK and follow it down to the bottom of the diagram to the number 400, indicating that the left side of the square is 400 kilometres east of the origin.

Now for the northing. Locate the bottom of the square SK and follow the line to the left to the number 300, indicating that the bottom of the square is 300 kilometres north of the origin.

The 400-kilometre easting crosses the 300-kilometre northing at the south-western (bottom-left) corner of the National Grid square SK. This point defines the reference of the square as 400 east and 300 north, written 400 300. The space between the numbers is not strictly necessary, and normally grid references are given without the space; in this case, as 400300.

The grid reference of the square is that of its bottom-left corner, with easting coming before northing.

Activity 11 Grid coordinates

Look at Figure 14(c). Which grid squares are defined by the following coordinates?

(a) 200 700 (b) 600 300

What are the grid references for the following grid squares?

(c) SY (d) TL

When planning a walk you would probably want to know the location of places to within 100 metres or so. Hence each 100-kilometre square is broken down into smaller squares. Figure 15 (overleaf) shows how this works. The square SK is divided up into smaller squares by drawing grid lines running east–west and north–south spaced at 10-kilometre intervals. This divides the large square into 100 squares.

Each such square is itself divided up into 100 smaller squares by drawing east–west and north–south grid lines at 1-kilometre intervals, as shown in the lower part of Figure 15. The grid reference of each of these smaller squares is the distance in kilometres measured towards the east and towards the north from the south-western corner of the large square. Remember that a grid reference is the reference of a *square*, not a single point.

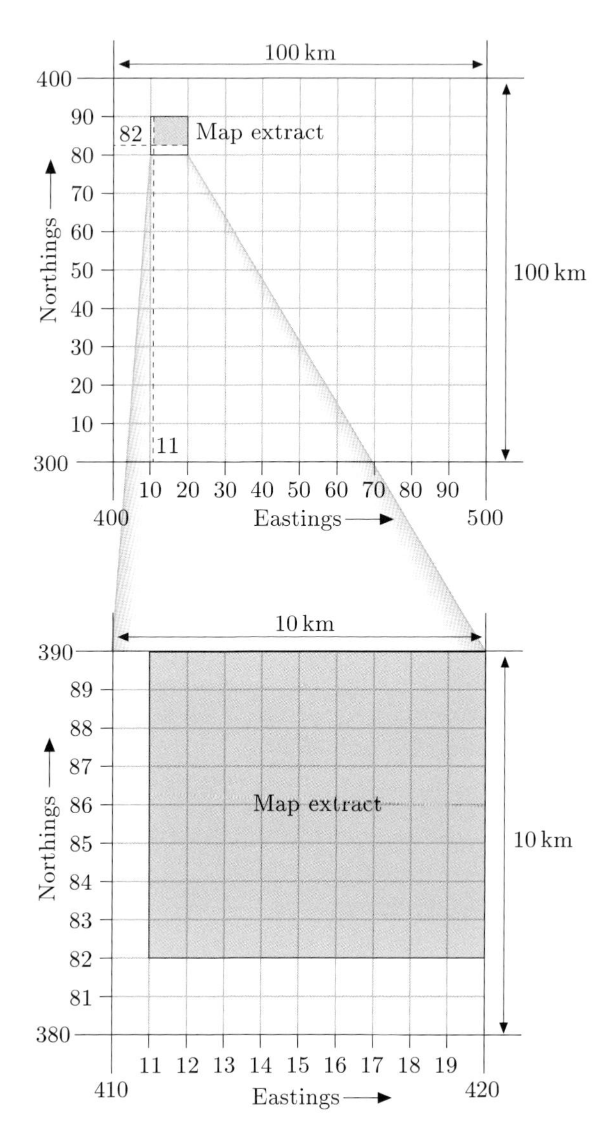

Figure 15 The location of your map extract in the OS national grid

If you look at the south-west corner of your map extract, you will see the easting $^{4}11^{000\text{m}}$ on the bottom edge, and the northing $^{3}82^{000\text{m}}$ on the left edge. Similar numbers are at other corners. The first numbers from each of these, 4 and 3 respectively, are in smaller type and give the reference (in hundreds of kilometres) of the SK square containing your map.

Remember the grid reference of the square is the coordinates of its bottom left-hand corner.

The numbers 11 and 82 printed in larger type give the distances in kilometres measured east and north from the south-western corner of the square SK, illustrated in Figure 15. On your map, these numbers go from 11 to 20 along the top and bottom, and from 82 to 90 up the sides. They label the blue grid lines spaced at 1-kilometre intervals.

Now you should be able to appreciate the extent of the land covered by your map extract. Your map extends 9 km east–west and 8 km north–south. And its south-western corner is itself 11 km to the east, and 82 km to the north of the south-western corner of the SK square.

The smaller numbers following the easting 11 and the northing 82 complete the reference, in metres. (Hence, the m at the end of the reference.) Thus, the south-western corner of your map is 411000 *metres* east and 382000 *metres* north of the origin of the National Grid. In practice, this degree of precision is not usually needed, and grid references are given within one square. So consider only the eastings and northings relevant to the square SK. The eastings are the numbers printed against the grid lines along the top and the bottom edges. The northings are the numbers printed alongside the grid lines up the left- and right-hand edges.

▶ Where do the grid lines with an easting of 13 and a northing of 84 cross?

The lines cross at the south-western corner of a 1-kilometre square which contains Hollins Cross, one of the viewpoints on the walk you will plan later.

However, the distances between the grid lines can be subdivided further into ten divisions, where each division corresponds to 100 metres. Estimating by eye or by using a ruler, you should find that Hollins Cross is just about six small divisions to the east of grid line 13, and just over five small divisions to the north of grid line 84, as shown in Figure 16. Hollins Cross, therefore, has an easting of 136 and a northing of 845, so the six-figure grid reference is 136 845.

This way of giving a map reference is the one used by the Ordnance Survey. But it is not the only way. You have come across others, such as the latitude and longitude system used in world atlases. Systems using combinations of letters and figures are used in street plans and road maps. Different conventions are used in different circumstances and it is important to know which one is being used in order to communicate information accurately and effectively.

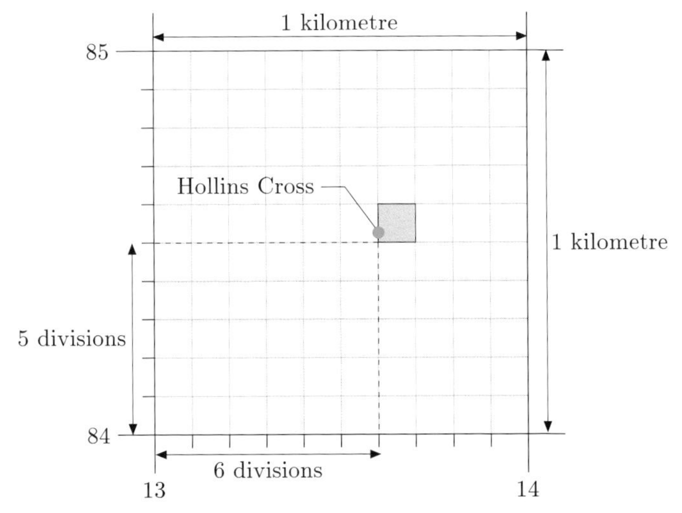

Figure 16 Grid reference of Hollins Cross

Recall that the easting is always given first and that the northing second and the grid reference usually written as a single string of digits: 136845. Since each coordinate has three digits, there should be no confusion about which numbers belong to the easting and which to the northing. The full grid reference of Hollins Cross is SK 136845. The letters SK indicate the large grid square in which Hollins Cross lies. However, the SK can be omitted when considering places in the same square.

Note the grid reference of the small square is again the coordinates of its bottom left-hand corner.

Where a location lies between two 100-metre divisions, the correct grid reference is given by the lower, not the nearer, of the two. Thus, grid references always round *down* and never round up.

When you are using grid references, bear in mind that a *six-figure reference refers to a 100-metre square* and not to a single point. All the points within the square share the same grid reference.

Locate Hollins Cross on your map and check the grid reference for yourself.

Example 2 Find the farm

Use your map to find the farm at 133840.

The grid reference is made up of an easting of 133 and a northing of 840. The 13 and 84 locate the south-western corner of the relevant 1 km square. Locate this square. Now by eye or using a ruler, mentally divide the lower side of the square into ten divisions. Count three to the right for 133. Since the northing is 840, there is no need to subdivide the northing. The south-west corner of the grid reference will be on the 84 northing. In this square is Mam Farm.

Activity 12 Using a grid reference

Use your map to find what is at the location given by the grid reference 127836.

Now let's look in detail at the route of the walk you are going to plan, using the grid references (given in brackets).

The walk starts at Mam Farm (133840) and follows the path to the south-west, shown as a green dotted line, changing to a black dotted line as it meets the route of an old road to the south, with Mam Tor lying to the west. The route follows the yellow road, which curves to the west below the Tor, eventually meeting the A625 road (128831). The route now follows the road to the west and joins a footpath (125831) near to a milepost, marked on the map as MP. The footpath goes north and then north-east to the top of Mam Tor (127836).

From Mam Tor, the walk follows the ridge to the north-east, passing Hollins Cross, Back Tor and on to Lose Hill. When the weather is clear this ridge offers spectacular views of Castleton and of the Vale of Edale to the north-west and beyond.

From Lose Hill, the route follows the path down to Losehill Farm (158846), where the walk finishes.

Activity 13 Finding grid references

Use your map to find the grid references of (a) Hollins Cross, (b) the point where the paths cross at Back Tor, (c) the view point on Lose Hill.

From a mathematical point of view, a grid reference is an example of using a pair of numbers (called coordinates) to specify a particular place on a two-dimensional surface such as a flat sheet of paper. The coordinates of a point on a graph are another example of this.

2.2 Symbols and scales

Your OS map represents information selected by the map-makers about the area around Castleton. It is not a picture, like Figure 17(a), but rather a structured collection of special symbols representing features, like Figure 17(b). A description of the symbols used on a map is given in a *legend* or *key*. The legend is at the bottom of your map sheet. It gives the symbols used to represent roads, paths, boundaries and rights of way, rocks, vegetation, camp sites, picnic sites, information centres, public conveniences, and so on.

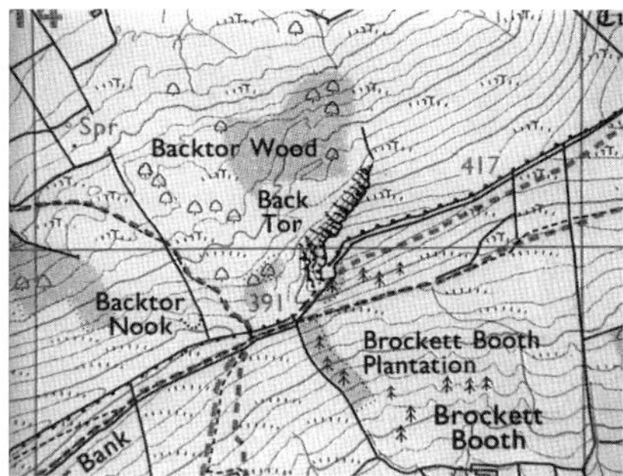

Figure 17 Two representations of the same feature: (a) a photograph, and (b) a map of Back Tor

The map has been compiled from large-scale surveys conducted between 1959 and 1979, and then revised several times more recently. Although the information included on the map is intended to be as accurate as possible, it reflects the revision and updating policies of the Ordnance Survey as much as reality itself.

Today, maps and charts of all types can be updated electronically using Geographical Information Systems (GIS)—see page 20.

Activity 14 Interpreting symbols

Locate and identify, using the legend on your map if necessary, the features specified by the following grid references.

(a) 172834 (b) 181832 (c) 123832 (d) 140866 (e) 187851

Scale

Thinking back to Section 1, the OS map preserves distances: equal distances on the map represent equal (horizontal) distances on the ground.

One important use of your OS map is to represent distance. The map uses a scale to relate distances measured on the ground to distances measured on the map.

Look at the map legend. At the top is printed:

Scale 1 : 25 000

This statement gives the scale of the map. It is written as a ratio, which means:

1 unit of distance measured on the map represents 25 000 of the same units measured on the ground.

Notice that it does not say whether the measurements should be in centimetres, metres, kilometres, miles, or some other unit. Because it is a *ratio*, it does not matter which units you use as long as you use the *same units* for both measurements.

If you are using centimetres, then 1 cm on the map represents 25 000 cm (250 m) on the ground. If you are using metres, then 1 m on the map represents 25 000 m (25 km) on the ground. Note that 'measured on the ground' means the horizontal ground distance between any two points.

To calculate the distance on the ground, multiply the distance on the map by the map scale. The relationship can be expressed as a word formula:

distance on the ground = map scale × distance on the map

where both distances are measured in the same units. For a 1 : 25 000 map such as yours, the formula is:

distance on the ground = 25 000 × distance on the map.

If the distance on the ground is D_G and the corresponding distance on the map is D_M, then the formula is

$$D_G = 25\,000 \times D_M.$$

You may like to add these formulas to your Handbook.

Rearranging the formula shows how map distances can be calculated from ground distances:

$$D_M = \frac{D_G}{25\,000}.$$

Remember that these relationships are valid only if D_M and D_G are expressed in the same units.

In everyday language, map scales are often expressed differently, as 'so many centimetres to a kilometre'. This allows you to estimate by eye the actual distances between places from a map.

To help you to estimate distances easily, your map shows two linear (straight-line) scales marked out directly in metres and kilometres in one case, and yards and miles in the other, showing what a scale of 1 : 25 000 means in everyday terms. If you use a ruler to measure the scales, you will find that there are 4 centimetres to a kilometre.

Example 3 How far is it on the ground?

On your map extract, the walk you are planning follows a path from Mam Farm (133840) to the A625 road (128831). It follows the road for 1.2 cm on the map. How far is this on the ground?

In the formula $D_{\mathrm{M}} = 1.2$ cm. So

$$\begin{aligned} D_{\mathrm{G}} &= 25\,000 \times 1.2 \text{ cm} \\ &= 30\,000 \text{ cm} \\ &= 300 \text{ m or } 0.3 \text{ km}. \end{aligned}$$

So the distance along the road is 300 m or 0.3 km.

Activity 15 Using a map scale

Assume a map scale of 1 : 25 000.

(a) A footpath between two villages is 12 cm long on the map. How far would it be to walk between the villages?

(b) A road between two places is 7.5 km long. What length would you find if you measured it on the map?

Estimating distances from a map is particularly useful in planning a walk. You want to know the distances from point to point along the route so that you can judge how long each section will take.

Activity 16 Calculating distances

Find the start of the walk at Mam Farm (133840) on your OS map. How far is it from there to the point where the route reaches the A625 road (128831)? The route is not a straight line, so you will have to estimate the map distance, perhaps using a piece of paper or string, or an opisometer (a small map-measuring device) if you have one.

Use the map scale to convert the map distance to ground distance in km.

The walk now follows the A625 road west until it meets a footpath at (125831) which it takes up to the top of Mam Tor (127836). The total distance from Mam Farm to Mam Tor on the map is about 8 cm. So

$$\begin{aligned} D_{\mathrm{G}} &= 8 \times 25\,000 \text{ cm} \\ &= 200\,000 \text{ cm} \\ &= 2 \text{ km}. \end{aligned}$$

This needs to be done for each stage of the walk. One way to organize the information about distances is by drawing up a table. Table 1 contains a partially completed list showing each stage with the corresponding map and ground distances.

Table 1 Table of distances for the walk

From (Grid reference)	To (Grid reference)	Measured map distance (in centimetres)	Distance calculated on ground (in kilometres)
Mam Farm (133840)	Mam Tor (127836)	8	2
Mam Tor (127836)	Hollins Cross (136845)	5.3	1.3
Hollins Cross (136845)	Back Tor (145849)	4.3	
Back Tor (145849)	Lose Hill (153853)	3.5	
Lose Hill (153853)	Losehill Farm (158846)		
Total distance			

Activity 17 Predicting distances

Complete Table 1 by measuring the distance on the map for the last leg of the walk and converting the map distances to ground distances in km.

How far is it in total from Mam Farm to Losehill Farm?

2.3 Look for the hills

Visitors to the Peak District are attracted by the magnificent tors (rocky outcrops), peaks and high moors: in short, by the hills. Look at your map of the Peak District now, to see where the hills are.

The contour lines on your map are drawn at 10-metre intervals. The lines join points that are the same height above sea level. On the east side of Mam Tor, for example, there are lines representing heights of 350 metres, 360 metres, 370 metres, and so on. Where the lines are well spaced, you should be able to see the height marked somewhere on the line. Where the lines are close together, such as near the top of the Tor, only a few labelling heights are printed. By counting in 10-metre steps from the nearest known contour, you can work out the height represented by an unmarked line. The contour lines at 50-metre intervals are slightly thicker. You can, perhaps, pick out the lines at 250, 300, 350, 400 and 450 metres on the east side of Mam Tor. Note that the contour heights are printed so that the tops of the numbers point towards the top of a slope.

The steepness of a slope is indicated by the spacing of the contour lines. If the ground is steep, the height changes rapidly over a small horizontal distance, so the contours will be close together. Where the ground is flatter the height changes more slowly, so the contour lines will be more widely spaced.

Figure 18 shows some examples of contour line patterns of concentric rings. Below each is the profile (or graph) of the hill it represents. The pattern represents a bird's eye view of a hill, with each contour line indicating the points around the hill that are at the same height.

On your map you can see a similar pattern of lines around Lose Hill (153853).

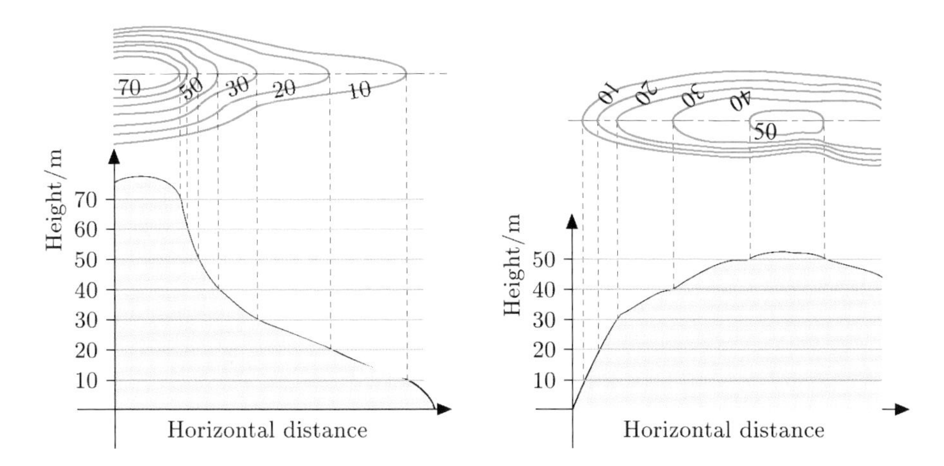

Figure 18 Different patterns of contour lines for different features

The spacing of the contours reflects different types of slope. Contours that bunch up indicate a slope that is getting steeper. In the left-hand hill in Figure 18, the bunching occurs near the top of a slope. In the right-hand pattern of contours, the bunching occurs near the bottom, indicating that the lower slopes are the steepest.

Imagine now that you are standing on the top of Mam Tor (127836). Look at the contours around this point on your map. The ground slopes away steeply on all sides except to the north-east, the direction of the path the walk will follow along the ridge to Hollins Cross. The height of the ridge path drops as you get closer to Hollins Cross, which is nearly 130 metres lower than the summit of Mam Tor, but since the height falls over a distance of about 1.3 kilometres, the slope of the path is nothing like as steep as the sides of the ridge.

On towards Back Tor, the path skirts the high point at 426 metres, marked as a spot height on your map, and follows the contours, reaching a point where several paths meet. At Back Tor itself, the contours are bunched together, indicating a sharp increase in the slope.

▶ If you could cut vertically downwards along the line of the path, what would the profile of the cut edge look like?

Figure 19 shows the profile of the path, from the Brocket Booth side and looking north-west towards the path. This section of the path will be a short but steep climb up to the top of Back Tor.

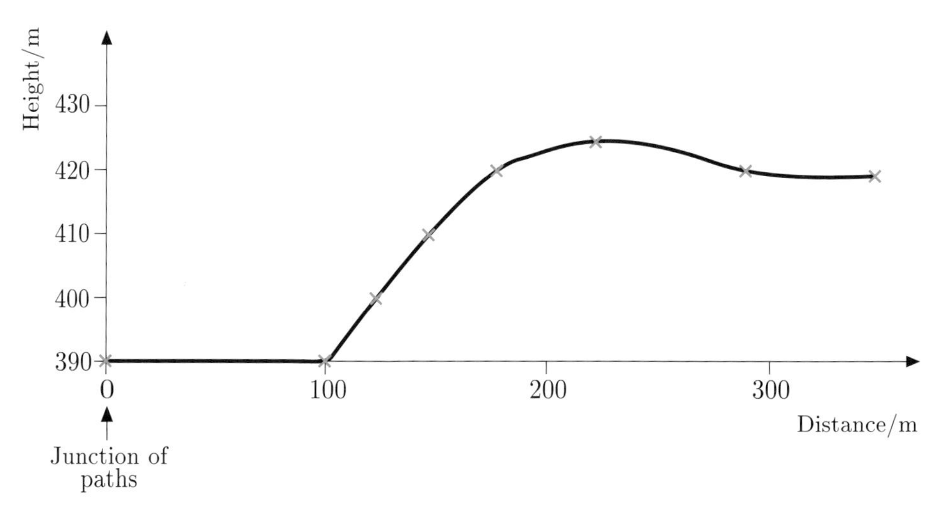

Figure 19 Profile of the path up Back Tor

Activity 18 Sketching profiles

Follow the path from Back Tor to Lose Hill on your map, noting where the contour lines cross the path, and what heights they represent. Make a sketch of the profile of the path.

2.4 Representing the walk

Now let's draw some ideas together. In Activity 17, you completed the table of distances of the different stages of the walk shown below.

Table 2 Distances along the walk

From (Grid reference)	To (Grid reference)	Measured map distance (in centimetres)	Distance cal on grou (in kilom
Mam Farm (133840)	Mam Tor (127836)	8	2
Mam Tor (127836)	Hollins Cross (136845)	5.3	1.3
Hollins Cross (136845)	Back Tor (145849)	4.3	1.1
Back Tor (145849)	Lose Hill (153853)	3.5	0.9
Lose Hill (153853)	Losehill Farm (158846)	3.4	0.9
Total distance		24.5	6.2

Using a table to record distances is one approach; drawing a profile is another. An alternative approach is to sketch a network map of the walk and write the distances on the arcs joining the vertices, stressing the main places along the walk and the distances between them. All other features of the route are ignored. The network map represents the path as a series of straight lines, without a scale or accurate orientation, as in Figure 20.

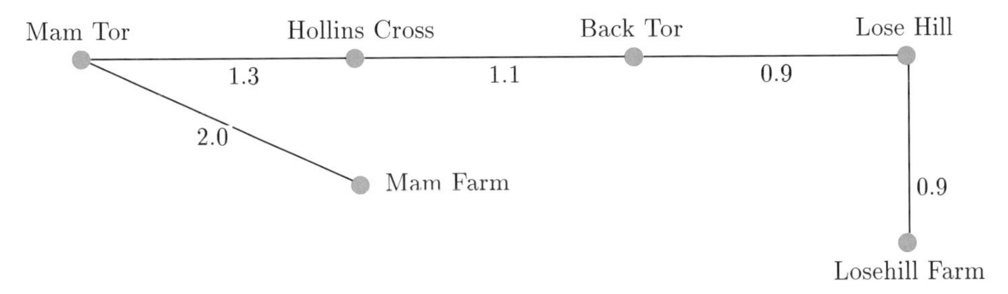

Figure 20 Network map of the walk, with distances in km

The table, the profile and the network map are different representations of the walk.

▶ Suggest advantages or disadvantages of the network representation.

An advantage of the network map is that it gives a simple visual impression of the walk, containing the main items of information: place names and distances. The simplicity is traded off against the more informative table and visual profile.

So you now know how far you will be walking. In the next section, you will think about directions and bearings, and how long the walk will take.

Before leaving this section, think about the techniques you have been using, such as using coordinates to locate positions precisely; using ratios and scales in the conversion between map distances and ground distances; interpreting height data. Summarize the main points, perhaps on your Handbook sheet, to help you remember and consolidate your learning.

A map is a symbolic representation of selected information. There are mathematical ideas embedded in your map, including a 2D (easting and northing) coordinate system for specifying position, scales to convert from one set of measurements to another, symbols to convey meaning, and contour lines to provide information about heights and slopes, from which graphical representations (profiles) can be drawn.

Outcomes

After studying this section, you should be able to:

◇ review your prior knowledge of a topic to help plan your studies (Activity 10);

◇ use grid references to specify locations (Activities 11 to 14);

◇ use the following terms accurately: 'grid system', 'map scale', 'legend', 'key', 'contour line', 'profile' (Activities 10 to 18);

◇ convert measurements of ground distance to their corresponding map distances and vice versa (Activities 15 to 17);

◇ explain how 2D contour line patterns represent 3D topography, and use them to sketch the profile of a walk (Activity 18).

3 Getting your bearings

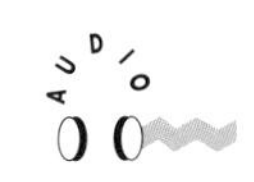

Aims This section aims to continue mathematical modelling in the context of maps, in order to help you develop an understanding of angles and direction bearings, and of how bearings can be used to specify a location. ◇

This section is in three parts. First, the video band 'Getting your bearings' shows you the route of the walk you planned in Section 2. It should help you visualize the 3D landscape from the 2D map and introduce the use of compass bearings. The video lasts about 25 minutes.

In Subsection 3.2, you will need your OS map extract and a protractor to measure angles. If you are unsure about using a protractor, listen to the audio band associated with the appendix.

Subsection 3.3 focuses on a rule of thumb, called Naismith's rule, used by walkers to estimate how long a particular walk will take. You will be writing down and using a formula for the rule. But also think about the assumptions Naismith's rule uses and the accuracy of its estimates.

3.1 Hill views

The video band shows the route of the walk through the eyes of three walkers. Compare the shape and features of the actual countryside with the image you may have in your mind's eye from the map. Not all the features the walkers come across are marked on the map, underlining the message that a map is a selective representation.

A different perspective is provided by the map-makers of the Ordnance Survey. The video also shows the Ordnance Survey computer-generated graphics relating the patterns of contour lines on maps to the heights, slopes and overall shape of the actual countryside. You will be able to see the path of the walk marked out on a three-dimensional view of the ridge from Mam Tor to Lose Hill, and be able to compare the geographical features of places such as Hollins Cross and Back Tor with the patterns of the corresponding contours.

The video reviews some of the topics you met in Section 2, but it also looks at the use of compass bearings to estimate position and direction.

You will probably find it useful to read through the next activity before you watch the video band and then complete the activity after viewing. As you watch, make notes and sketches for the activity.

Now watch band 5 on DVD 00107 called 'Getting your bearings'.

Activity 19 *Using the video*

Imagine you have been asked by the local scout leader to help a small group of scouts prepare for the walk shown in the video. Prepare notes to teach the scouts:

(a) how to take a compass bearing;

(b) to recognize particular features from the map contours, e.g. a saddle point.

3.2 Measuring and predicting bearings

On the walk you use a compass to check your bearings and navigate. A compass bearing is the direction, given by a magnetic compass, of one point from another (using the conventions described in Section 1, page 10).

Another convention in Ordnance Survey maps is that the northerly direction of the vertical grid lines on your map is called 'grid north'. Because an OS map is drawn as a flat representation of a curved surface, grid north differs slightly from 'true' north, i.e. the direction of the North Pole. On your map, the difference is very small.

However, there is yet another north which is important to anyone using a magnetic compass. The north-seeking needle of a magnetic compass points to magnetic north, the geomagnetic pole in the Northern Hemisphere.

It is important to be clear about the difference between grid north and magnetic north. Grid north is a direction that does not change with time. Magnetic north, on the other hand, is a physical, changing phenomenon owing to gradual changes of the Earth's magnetic field.

Records of the difference between true north and magnetic north have been kept for England since the late sixteenth century. In London in 1576, magnetic north was estimated to be over 10° *east* of true north. Over the next 250 years, it drifted west by more than 30°, reaching a maximum of nearly 25° west in 1823. It is now drifting back eastward.

In the legend on your map, you will find some information about how grid north and magnetic north are related. In 1994, when the video was filmed, the direction of magnetic north was about 5° west of grid north.

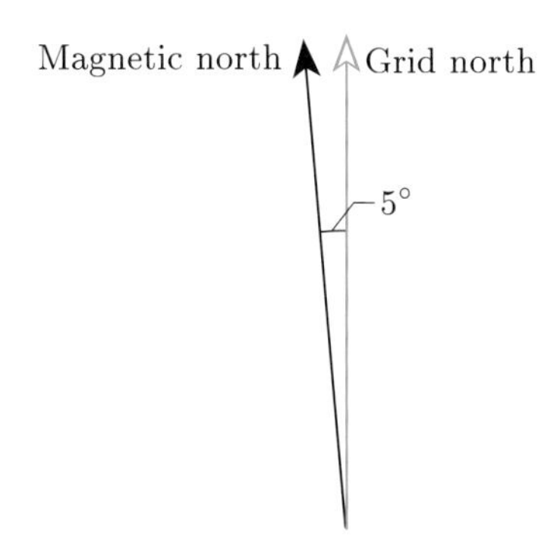

Figure 21 Magnetic north and grid north

In 1994, the difference between the direction of magnetic north and grid north was decreasing at a rate of about a sixth of a degree per year.

This information can be represented by a formula which can be used to predict the direction of magnetic north in the future. In the formula, the number of degrees that magnetic north is to the west of grid north is M and the number of years after 1994 is Y. Then

$$M = 5 - \frac{Y}{6}.$$

This information is valid only for the region covered by your particular map. The direction of magnetic north varies with place as well as with time. So on a map of a different part of Great Britain or in another country, the angle between grid north and magnetic north may be different.

Example 4 Using a formula

Use the formula to predict the direction of magnetic north in 2006 (12 years after 1994).

In 2006, $Y = 12$. So $M = 5 - \frac{12}{6} = 5 - 2 = 3$.

So the formula predicts that magnetic north will be 3° west of grid north in 2006.

Activity 20 Predicting magnetic north

(a) Use the formula to predict the direction of magnetic north in 2012 for the region covered by your map.

(b) The direction of magnetic north is currently drifting towards grid north. If the current rate of drift is maintained, in what year will the directions of magnetic north and grid north coincide for your map?

Recall, from Section 1, that the OS map preserves distances and directions.

Take stock for a moment. You have an OS map on which a grid has been overlaid. This grid has been defined by the map-makers and anyone can use it to fix locations and to measure angles between different points on the map. A bearing measured on a map relative to grid north is called a grid bearing.

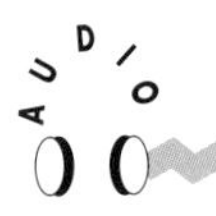

If you are unsure how to use a protractor to measure angles up to 360°, work through audio band 1 of CD5509 along with the frames in the Appendix on page 68.

Example 5 Bearing of A from B

From Mam Tor, you can see the village of Nether Booth (grid reference 142861) in the Vale of Edale. Find the grid bearing of Nether Booth from Mam Tor.

Using a pencil and ruler, draw a straight line on your map connecting Mam Tor and Nether Booth. Now draw a line through Mam Tor in the direction of grid north, parallel to the north–south grid lines. Figure 22 is a sketch of the bearing. Using a protractor, measure the angle between grid north and the line to Nether Booth.

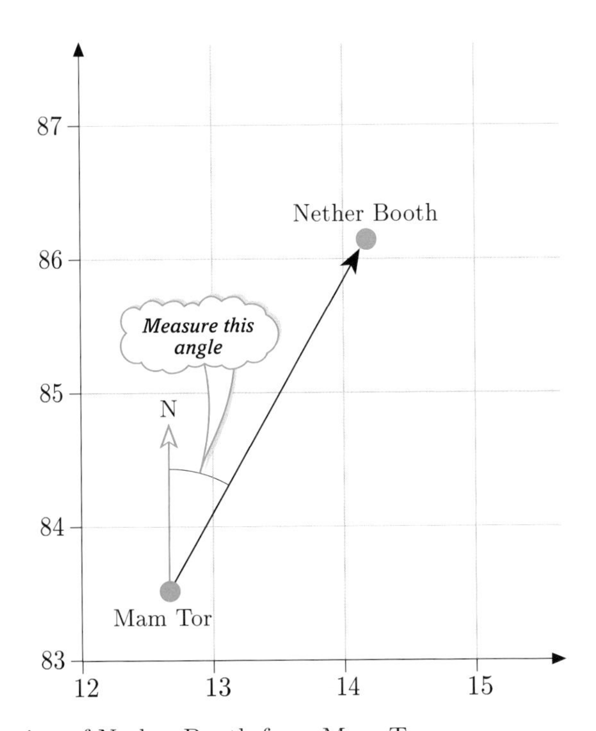

Figure 22 Bearing of Nether Booth from Mam Tor

You should have found that the grid bearing of Nether Booth from Mam Tor is about 30°.

▶ Now what would happen if you actually stood on Mam Tor and used a compass to take a bearing on Nether Booth?

The reading would be more than 30°. Why? Because the compass bearing gives the angle between the direction of *magnetic north* and Nether Booth. Since magnetic north lies to the west of grid north, the compass bearing is *greater* than you measured on the map. Figure 23 (overleaf) shows how this works for 1994, when the video was made.

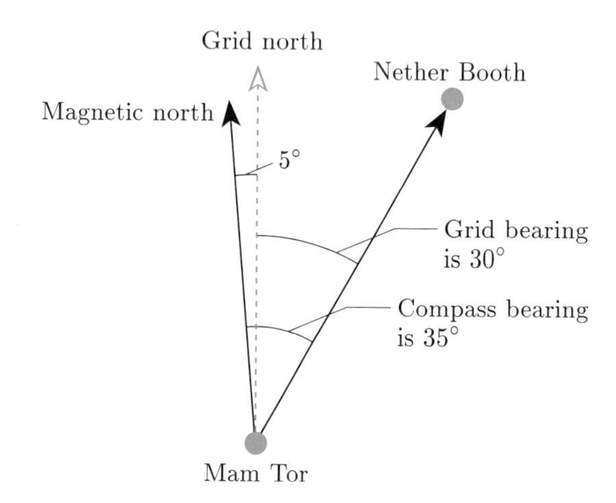

Figure 23 Grid and compass bearing of Nether Booth from Mam Tor

So there is something to remember when using a compass and an OS map: to convert grid bearings to compass bearings, add the difference between grid north and magnetic north.

Of course, this rule will have to change as magnetic north moves to the east of grid north.

To convert compass bearings to grid bearings, subtract the difference.

Among walkers there are several mnemonics used to help remember whether to add or subtract. These include 'add for mag, get rid for grid'.

Activity 21 Checking directions

On the video, you saw the walkers (in 1994) check the direction of the path at a point (grid reference 132839) near Mam Farm. Use your map to find the grid bearing and hence predict the (1994) compass bearing from this point to the junction of the path and the road (grid reference 128831).

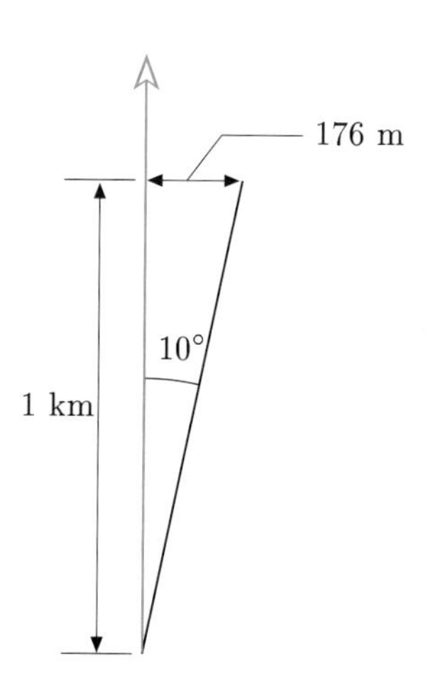

Figure 24

On walks in clear weather, where a compass is used to take a bearing on a landmark such as a hilltop, the difference between grid north and magnetic north is unlikely to cause any major problems in navigation. However, if you are crossing open country in the mist, accurate bearings are crucial for your safety and it is important to take the difference into account. Errors can be compounded, so that a walker might end up following a bearing that may be out by 10°. Over a kilometre, a 10° error would result in a deviation of over 176 m from the required path.

Activity 22 Predicting compass bearings

Suppose you were on a walk in the High Peak on Blackden Moor in 1994, and wanted to check your position when you reached the junction of paths at grid reference 114888, north of Seal Stones.

Using your map, predict the compass bearings of Upper House Farm (119899) across the valley and of Blackden View Farm (132896), further to the east, from your supposed location 114888.

Now think about Activity 22 the other way around. What would be the compass bearing of the location 114888 from each farm? Figure 25(a) shows that the reverse bearing is the original bearing, θ (theta) on the diagram, plus 180°, i.e. $\theta + 180°$. So looking back from Upper House Farm $\theta = 31°$ and so the compass bearing will be $31° + 180° = 211°$ (see Figure 25(b)).

This is true whether you are using compass bearings or grid bearings.

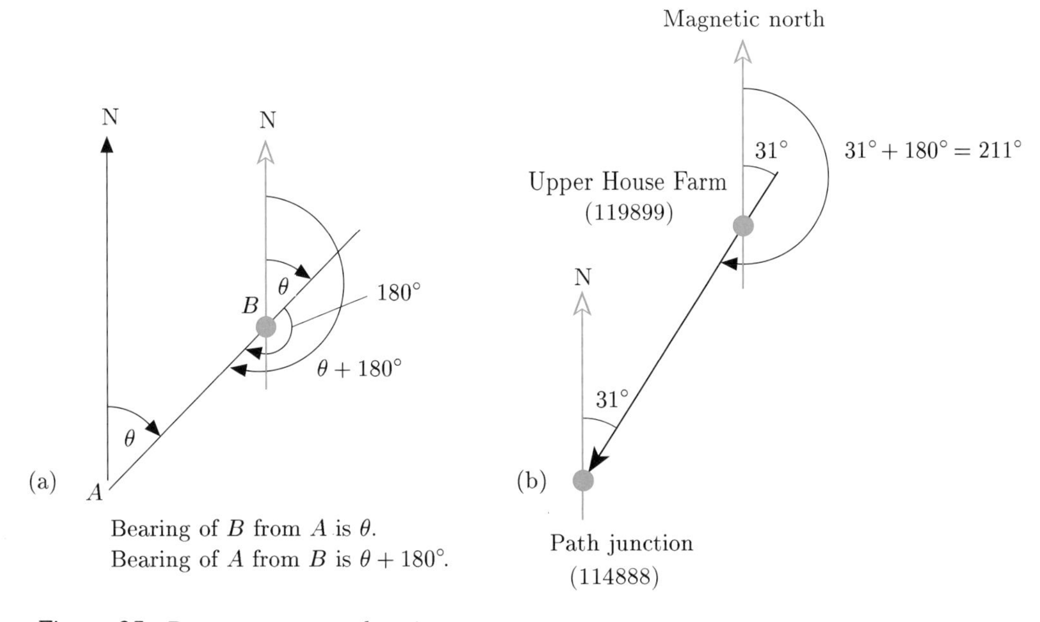

Figure 25 Reverse compass bearings

Activity 23 Bearing from Blackden View Farm

From Activity 22, you saw that the bearing of Blackden View Farm from the junction of paths was 72°. What would the bearing of the junction of paths be from Blackden View Farm?

Two bearings are enough to fix your location on Blackden Moor. But what if the bearings you measure are not the ones you expect?

Finding the position of a third point based on the bearings taken from two others forms the basis of a technique called *triangulation*, used in surveying.

Activity 24 Where am I?

Suppose you got compass bearings of 18° for Upper House Farm and 61° for Blackden View Farm (in 1994). Where were you?

So far, all the reverse compass bearings were less than 360°. But what would it mean if the result of adding 180° to a bearing came to more than 360°? Suppose you started with a bearing of 230°. Adding 180° to this gives 410°. But a bearing of 360° is the same as a bearing of 0°, so 410° corresponds to $410° - 360° = 50°$.

A map is a 2D surface. Any point on a map can be located by just two independent items of information. On an OS map, the items of information may be a pair of bearings, or an easting and a northing. On a world map, the information may be latitude and longitude.

To finish this subsection, return to the walk. On the video you saw the walkers use their compass at Lose Hill to check the direction of the path to Losehill Farm.

Activity 25 Journey's end

Using your OS map and a protractor, predict the compass bearing of Losehill Farm from the top of Lose Hill (in 1994).

3.3 Time for a walk

It is useful to be able to estimate the time that a walk will take. A well-known rule of thumb, called Naismith's rule, is used by walkers to estimate the time it will take to walk across open, hilly country.

Naismith's rule has two parts: one is concerned with an average walking speed over flat ground and the other with the extra time it takes to climb up any slopes. In the modern form of the rule, the walking speed is expressed in kilometres per hour and the height of ascent in metres. It can be stated as follows.

The time required for a walk in open hilly country can be estimated by:

(a) allowing an hour for every (horizontal) 5 km walked;

(b) adding an hour for every (vertical) 600 m climbed.

Notice that Naismith's rule does not include any extra time for coming down a slope, regardless of how steep it is.

> ### *Historical note*
>
> William Naismith (1856–1935) was a Scottish climber and alpinist. He formulated his rule for estimating hill walking times in 1892. The original rule assumed a walking speed of three miles per hour, plus an extra thirty minutes for every thousand feet of ascent.

To estimate the time to complete a walk, you need to know how long the walk is in kilometres, and how much height you expect to ascend (adding up all ascents, not just looking at total height gained overall).

▶ Stop for a moment and think about how you would write Naismith's rule as a formula. How would you complete the following sentence?

The estimated time for a walk is equal to . . . . plus . . . .

Naismith's rule can be split into: a horizontal walking part (covering the horizontal distance at 5 km per hour) and a vertical climbing part (covering the vertical distance at 600 m per hour).

The horizontal walking part

Since it takes one hour for every 5 km, the walking time in hours is equal to the distance (D) in kilometres divided by 5.

Or, more concisely,

$$\text{walking time (in hours)} = \frac{\text{distance in kilometres}}{5} = \frac{D}{5}.$$

For example, for a 10 km walk over level ground, $D = 10$. So the estimated time is $10/5 = 2$ (hours).

The vertical climbing part

Note first how the total ascent is calculated. Figure 26 shows the profile of a path that starts at a height of 50 m, climbs to 80 m and then descends to 40 m before climbing again to 110 m. Although the overall change in height from 50 m to 110 m is 60 m, Naismith's rule requires the individual ascents to be added together, giving a total of $30 + 70 = 100$ metres.

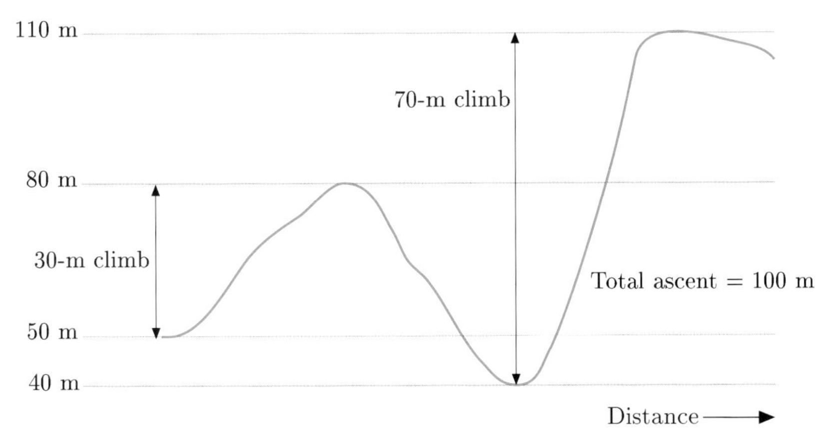

Figure 26 Calculating height for Naismith's rule

So a climb of 10 metres, from one contour line to the next on an OS map, adds about 1 minute to the estimated walking time.

The assumption is that an ascent of 600 m will add an extra hour to the time. So the extra climbing time in hours is equal to the total ascent (H) in metres divided by 600. Or, more concisely,

$$\text{extra climbing time (in hours)} = \frac{\text{total ascent in metres}}{600} = \frac{H}{600}.$$

So, if a walk involves a total climb of 300 m, it will add about $300/600 = 0.5$ hours, or 30 minutes, to the walking time.

Naismith's rule puts the walking time and the extra climbing time together in one formula to give an estimate in hours of the total time for a walk.

$$\text{Time taken for walk} = \text{horizontal walking time} + \text{vertical climbing time.}$$

Or,

$$\text{time taken for walk in hours} = \frac{\text{distance in kilometres}}{5} + \frac{\text{total ascent in metres}}{600}$$

Denoting the total time in hours by T gives:

$$T = \frac{D}{5} + \frac{H}{600}.$$

Notice that both parts of this formula have hours as the units of time, even though the unit of length is kilometres in the first part, and metres in the second part. This is because the average walking speed is expressed in *kilometres* per hour, while the average climbing speed is expressed in *metres* per hour. Dividing distance in kilometres by an average walking speed in kilometres per hour gives an answer in hours, as does dividing the total ascent in metres by an average climbing speed in metres per hour.

This is an important point to appreciate. Physical quantities such as times or distances or speeds are not just numbers; they are always associated with units. When such quantities are added or subtracted, the units should be the same for each quantity. It is meaningless, for example, to add 5 km to 100 m to get 105. The sum makes sense only if both are expressed in the same units: 5 km + 0.1 km = 5.1 km, or 5000 m + 100 m = 5100 m.

Example 6 A long walk

Use Naismith's rule to estimate the time taken for a 20 km walk that involves a total climb of 1200 m.

$D = 20$ and $H = 1200$.

Naismith's rule is $T = \frac{D}{5} + \frac{H}{600}$. Replacing D by 20 and H by 1200 gives:

$$T = \frac{20}{5} + \frac{1200}{600}.$$

Doing the division gives

$$T = 4 + 2 = 6.$$

So Naismith's rule predicts a 6-hour walk (not including breaks!).

Activity 26 Using Naismith's rule

The first part of the walk from Mam Farm to the summit of Mam Tor is about 2 km long, and the ground rises from about 300 m to 517 m.

Use Naismith's rule to estimate how long this section will take.

Naismith's rule is a *model*, a formulation about walking time in general giving predictions for future walks. It is a *mathematical* model because it uses mathematics to predict the time, instead of doing the walk to find out.

You will come across other mathematical models in this course. All such models use assumptions about the situation they represent. They stress some features and ignore others. Stop for a moment and think about the assumptions underlying Naismith's rule.

Activity 27 What are some assumptions?

Think carefully about Naismith's rule.

◇ What do you think Naismith's rule stresses and what does it ignore about hill walking? Note down as many ideas as you can think of.

◇ Where do you think the figures for walking and climbing speed come from?

Now write some notes on Naismith's rule, including the formula, for your Handbook.

Naismith's rule is based on experience. The average walking and climbing speeds (of 5 km and 600 m per hour) have been chosen so that the rule gives a fair estimate of time for reasonably experienced hill walkers. Naismith's rule is less likely to give reasonable predictions for a group of young children, or for Army paratroopers on an exercise march.

The average walking and climbing speeds used in Naismith's rule are called the *parameters* of the model. The values of the parameters could be changed for other groups of people.

Thinking about the assumptions underlying a model is important when you are applying the model. Sometimes the assumptions do not fit a particular situation, and you may have to make some adjustments.

To finish this subsection, use Naismith's rule to estimate the time for the whole walk from Mam Farm to Losehill Farm, but be wise and allow a safety margin in your time estimates. If you were just about to set out on this walk it would be a good idea to gather all the information about places, grid references, distances, compass bearings and times together on a handy reference route card. You can take a route card with you and leave a copy to let others know where you were planning to go, and your time estimates. Figure 27 shows a route card partially completed for this walk.

Date	11 April 1995				
From: Mam Farm		grid ref: 133840		starting time: 10 am	
To: Losehill Farm		grid ref: 158846		est. arrival time:	
Path to	Grid reference	Initial Compass bearing	Distance (kilometres)	Height climbed (metres)	Estimated time (minutes)
A625 road	128831	212°			
Footpath	125831	287°	2	217	50
Mam Tor	127836	352°			
Hollins Cross	136845	24°	1.3	0	20
Back Tor	145849	91°	1.1	40	20
Lose Hill	153853	65°	0.9		
Losehill Farm	158846	156°	0.9		
Total			6.2		

Figure 27 A route card for the walk

The fourth column gives the distances in kilometres for each stage and the fifth gives the total ascent in metres. From these the estimated time for each leg of the walk is calculated using Naismith's rule, with appropriate allowances for short stops, and entered in column 6.

Naismith's rule gives estimates. It makes sense to round answers up to the nearest 5 or even 10 minutes.

Here are some sample calculations. The distance from Mam Tor to Hollins Cross is 1.3 km, so $D = 1.3$. There is no climb to take into account because the path drops all the way, so $H = 0$. Naismith's rule gives the estimated time as

$$T = \frac{D}{5} + \frac{H}{600} = \frac{1.3}{5} + \frac{0}{600} = \frac{1.3}{5} = 0.26.$$

This is 0.26 hours. To change this to minutes multiply by 60: $0.26 \times 60 = 15.16$, but round up to 20 minutes for time to admire the views.

Back Tor is 1.1 km on from Hollins Cross, so $D = 1.1$ for this stage. The path rises about 10 metres to follow the 400-metre contour, falling back to a height of 391 metres as it approaches Back Tor. There is then a sharp rise of about 30 metres up the side of the Tor, so $H = 10 + 30 = 40$. According to Naismith's rule the time taken to walk to Back Tor from Hollins Cross will be

$$T = \frac{D}{5} + \frac{H}{600} = \frac{1.1}{5} + \frac{40}{600} \simeq 0.22 + 0.07 = 0.29.$$

In minutes this is $0.29 \times 60 = 17$ (to the nearest whole number). Again, round up to 20 minutes.

Activity 28 Completing the route card

Now complete the route card for the last two stages of the walk.

Use your OS map to find the height data for the last two stages of the walk. Then use Naismith's rule to predict the time they will take.

Estimate the total time for the walk, by adding up the stage times.

Check your estimate by adding the distance and heights climbed for the stages and then using Naismith's rule on the complete walk.

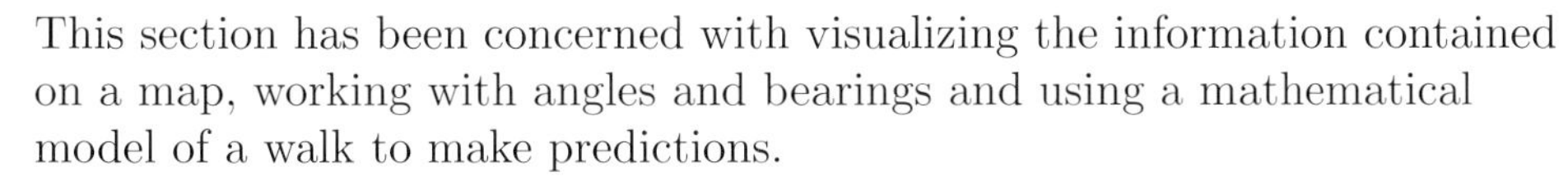

This section has been concerned with visualizing the information contained on a map, working with angles and bearings and using a mathematical model of a walk to make predictions.

An understanding of the map and compass conventions used were necessary to interpret the information correctly. The video showed different perspectives of the planned walk. One perspective was that of the walkers with the help of the Ordnance Survey map and a compass.

Another perspective was provided by computer-generated graphics. Using the information given by the contour lines, a three-dimensional representation of the landscape was built up. From this view you were able to see how the patterns of contour lines are interpreted in terms of the shapes of the hills and valleys.

Subsection 3.2 looked at angles and direction bearings. In map work (although not mathematics), angles are measured in degrees clockwise from north. A correction is needed for the difference between magnetic north and grid north.

Bearings taken on landmarks can be used to check position. On a two-dimensional surface using a fixed reference system, such as a map, two numbers are needed to specify a given location. These numbers may be an easting and northing pair (on the National Grid system) or two compass bearings.

Subsection 3.3 discussed Naismith's rule for estimating the time for a walk. It expresses (in words or symbols) a mathematical relationship between the distance walked (D km), the height climbed (H m) and the time taken for the walk (T hours). It is a simple mathematical model based upon assumptions about walking speed. The interpretation of the numerical result needs to take this into account.

Before completing this section, take a few minutes to review your work and your Handbook entries and the outcomes below.

Do you remember the model of learning that was described in *Unit 1*? Reviewing your learning can be thought of as the continual cycling back and forth between experience (the learning you have been involved in) and reflection.

Outcomes

After studying this section, you should be able to:

◇ make notes from a video, with the aim of giving an accurate description to others (Activity 19);

◇ appreciate that a mathematical model stresses some features and ignores others (Activity 27);

◇ use the following terms accurately and be able to explain them to someone not taking this course: 'grid bearing', 'compass bearing', 'grid north', 'magnetic north', 'mathematical model' (Activities 19–28);

◇ use a protractor to measure grid bearings on an OS map (Activities 21, 22 and 25);

◇ predict a compass bearing given a grid bearing and information about magnetic north, and vice versa (Activities 21, 22 and 25);

◇ find the reverse compass bearings (Activity 23);

◇ fix a location using the bearings from two other points (Activity 25);

◇ use a formula, such as that for Naismith's rule (Activities 26 and 28).

4 Slopes and sizes

Aims The main aim of this section is to explore slopes and areas, using formulas and graphs. ◇

The spacing of contour lines on an Ordnance Survey map gives an indication of the steepness of the ground. However, you can be more precise about the slope of a hillside, or the gradient of a road, using height and distance information from a map. This gives a numerical representation of gradient.

An OS map can be used to predict the size of areas as well as distances. Areas on the map are representations of areas on the ground. An area on the map is related to the corresponding area on the ground by a scaling factor. The area scaling factor is not stated explicitly, but it can be calculated from the map scale for *lengths*.

Thinking back to Section 1, the OS map preserves true area.

4.1 Picturing steepness

The steepness of a slope is called its *gradient*. From a mathematical point of view, the gradient expresses a relationship between change in height and horizontal distance. The average gradient of a slope between two points is defined as the increase in height divided by the horizontal distance travelled, as shown in Figure 28.

If you are working from a map, the increase in height can be found from the contour lines (and/or spot heights). The horizontal distance is found by multiplying the map distance by the map scale.

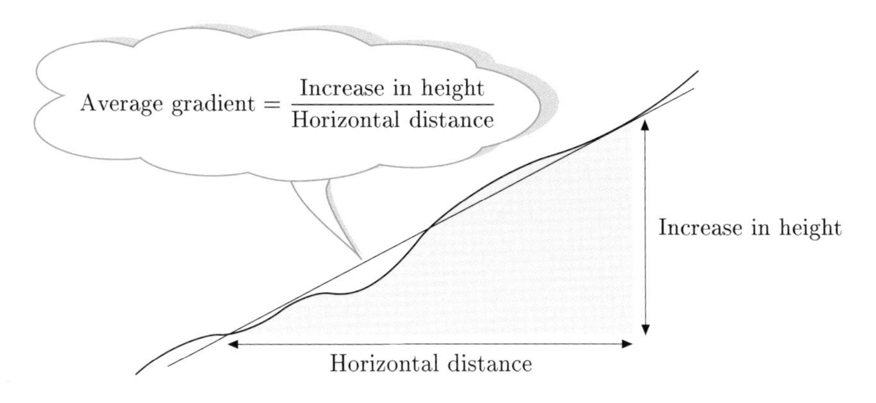

Figure 28 Definition of average gradient

So the average gradient is defined as:

$$\text{gradient} = \frac{\text{increase in height}}{\text{horizontal distance travelled}}.$$

If H is the increase in height and D is the horizontal distance travelled, then the formula for the gradient G is

$$G = \frac{H}{D}.$$

Here is an example from the walk. The path rises about 66 metres over a distance of roughly 500 metres as it climbs to the top of Lose Hill.

▶ How steep is the path?

In this case, $D = 500$ m and $H = 66$ m. So the gradient is given by

$$G = \frac{H}{D} = \frac{66 \text{ m}}{500 \text{ m}} = 0.13 \text{ (2 d.p).}$$

This is interpreted to mean that the ground rises 0.13 metres (13 cm) for every metre moved horizontally: a reasonable slope to walk up.

If both the increase in height and the horizontal distance are measured in metres, the gradient relates metres measured vertically to metres measured horizontally. But what would be the gradient if you measured height and distance in some other units?

▶ Would the gradient be a different number if you measured in centimetres or kilometres?

Activity 29 Finding gradients

On the north-west slope leading up to Hollins Cross, the ground rises by 100 m over a horizontal distance of about 300 m.

(a) What is the gradient of the slope, using measurements in metres?

(b) Convert the measurement to centimetres by multiplying by 100. Calculate the gradient using measurements in centimetres.

(c) Convert the measurements to kilometres by dividing by 1000. Calculate the gradient using measurements in kilometres.

(d) Explain why your answers are all the same.

The gradient is a ratio of the height to the distance, so as long as both measurements are made in the same units, their ratio will be a number (with no units) and the same, irrespective of the units used.

If the path slopes downwards then the "increase in height" is taken as negative. The convention is to think of the decrease as a negative increase, in the same way as taking money out of your bank account could be thought of as negatively increasing your bank balance. Here is an example from the walk. The map shows that the path from Back Tor dips from about 430 to 410 metres over about 300 metres from the top of Back Tor. So "increase in height" is $H = 410 - 430$ metres $= {}^{-}20$ m. The horizontal distance is $D = 300$ m. So the gradient is given by

$$G = \frac{H}{D} = \frac{{}^{-}20}{300} = {}^{-}0.067.$$

This is a small negative (downhill) gradient. Negative gradients represent going downhill.

Activity 30 A steep slope

Suppose you were on the top of Mam Tor. The slope to the east is very steep indeed, too steep for a path. It drops by 150 m over 200 m.

What is the average gradient of the slope on this side of the Tor?

Remember that the gradient is the ratio of a vertical distance to a horizontal distance. The gradient is simply a number. For a particular slope, the number is the same irrespective of the units.

▶ Can you give another example where a pure number is used to relate two measurements of length?

Recall map scales from Subsection 2.2. The relationship was:

$$\text{distance on the ground} = \text{map scale} \times \text{distance on the map}$$

where the map scale is a number without any units, and the distances on the ground and on the map are measured in the same units as each other.

Like gradient, map scale can also be thought of as a ratio:

$$\text{map scale} = \frac{\text{distance on the ground}}{\text{distance on the map}}.$$

This formulation emphasizes that it is a pure number with no units.

The relationship between height (H) and distance (D) can be written in a similar way to that of map scale:

$$\text{increase in height} = \text{gradient} \times \text{horizontal distance}$$

or

$$H = G \times D$$

where the increase in height and the horizontal distance are in the same units, and the gradient G is simply a number.

4.2 Drawing the slope

Now return to the path from Back Tor to Lose Hill. Find Back Tor on your OS map. Recall that at the top of Back Tor the path reaches a height of almost 430 metres, dipping to about 410 metres before climbing Lose Hill.

One way to get a feel for gradient is to make a sketch graph showing how the height of the ground changes as you move away from a particular point. By measuring along the line of the path on the map from Back Tor, you can estimate the distances from the Tor to where the path cuts the contour lines at 410 m, 420 m, 430 m etc., to the 476 m spot height at the top of Lose Hill. One person's measurements are given in Table 3 (overleaf).

Table 3 Distances and heights from Back Tor to Lose Hill

Distance from Back Tor (in metres)	Height (in metres)
0	420
83	430
170	420
370	410
500	420
600	430
670	440
720	450
750	460
820	470
870	476

Plotting this information as a sketch graph gives the profile of the walk: a representation of how the height changes with distance. Figure 29 has some of the data plotted.

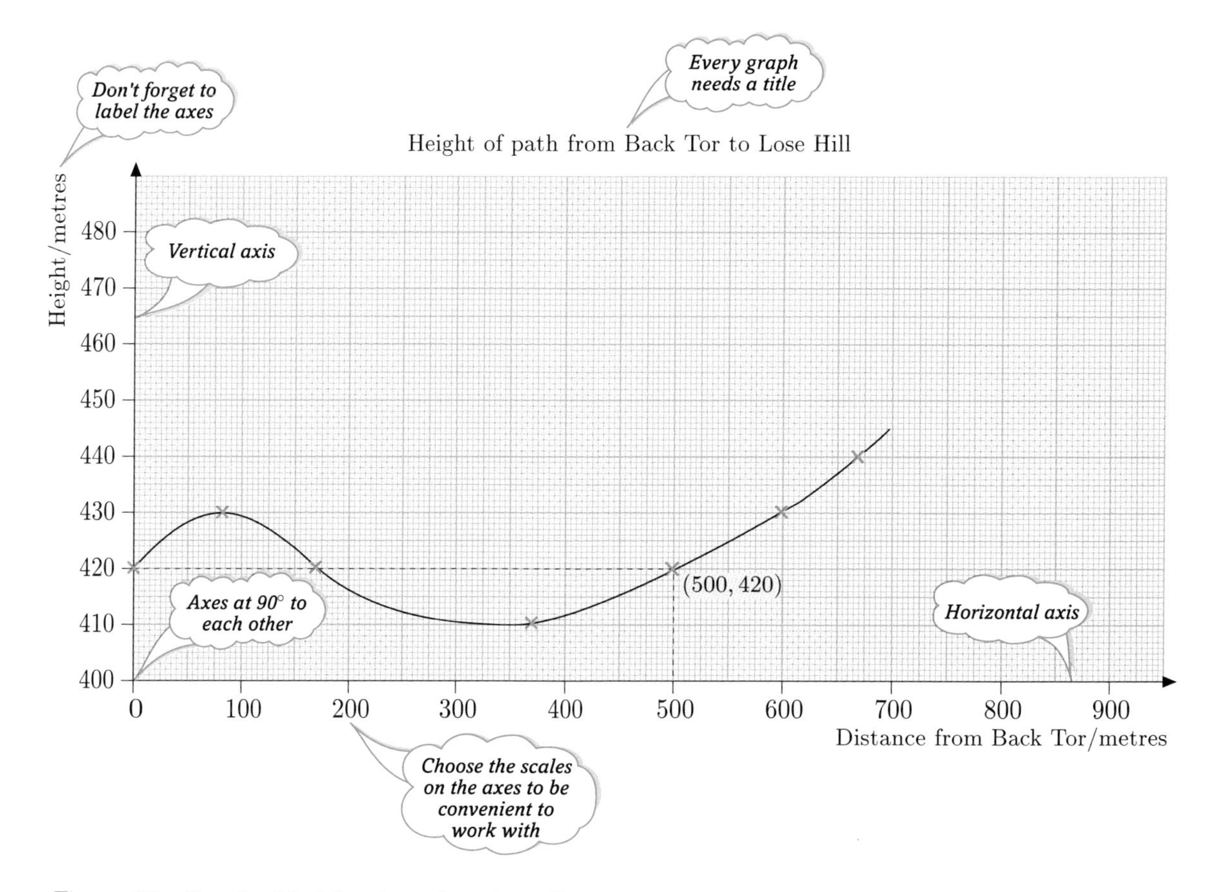

Figure 29 Graph of height plotted against distance

Note some mathematics conventions about graphs that are illustrated here.

The frame of the graph is provided by two *axes*: the horizontal axis running across the page, and the vertical axis running up the page. The angle between the axes is 90 degrees; mathematically they are said to be *perpendicular* to each other.

The word 'axes' is the plural of axis.

The horizontal axis shows the scale 0, 100, 200, 300, ..., 900 spaced at equal intervals. The axis is labelled to indicate that it represents distance in metres along the path from Back Tor.

The vertical axis shows the scale 400, 410, ..., 480. The label on this axis indicates that it represents the height of the ground in metres.

A practical point to note here is that when you choose the scale for the axes of a graph, choose a scale that is convenient and easy to use with your data. In Table 3, most heights are listed every 10 m, so plotting is easier using a vertical scale graduated in intervals of 10 m as in Figure 29.

Note that the graph has a title, so everybody knows what it represents.

Each point on the graph comes from a pair of numbers. The first number is the distance along the path, and the second number is the height of the ground. For example, in the table, at a distance of 500 metres from Back Tor the height of the path is 420 metres. By convention in mathematics, the coordinate on the horizontal or x-axis is written first. So this point is written as (500, 420). The numbers 500 and 420 are called the *coordinates* of the point, and the notation (500, 420) is called a *coordinate pair*.

This is not quite the same convention as that used for quoting OS grid references. Here the coordinates are given in the same order, but without the comma and brackets because the easting and the northing always have the same number of digits. But in mathematics, the two coordinates may well not have the same number of digits, so the comma is used to separate them. Also, in mathematics, a coordinate pair refers to a specific *point* on a graph, and not to a particular *square region*, as in OS grid references.

The small crosses are joined by a smooth line producing a *graph*. This is a *model* representing the profile of the ground. It assumes the height of the ground changes smoothly from one discrete coordinate point to the next.

Activity 31 Plotting points

Plot and join up, using a smooth curve, the remaining points from Table 3 on Figure 29 to complete the graph of height against distance.

Using the graph, how would you describe in words how the height of the ground changes as you move from Back Tor to Lose Hill?

Recall, from page 50, the figure of 0.13 for the gradient of the path up Lose Hill, based on an overall height rise of 66 metres over a distance of 500 metres. This is the average gradient of the slope, shown on Figure 30 by the dashed line. It stresses the increase in height and the horizontal distance, but it ignores all other details. It does not predict the slope of any particular part of the path, but rather gives an expectation about its general degree of steepness. At some points the gradient is steeper than average and at others less steep.

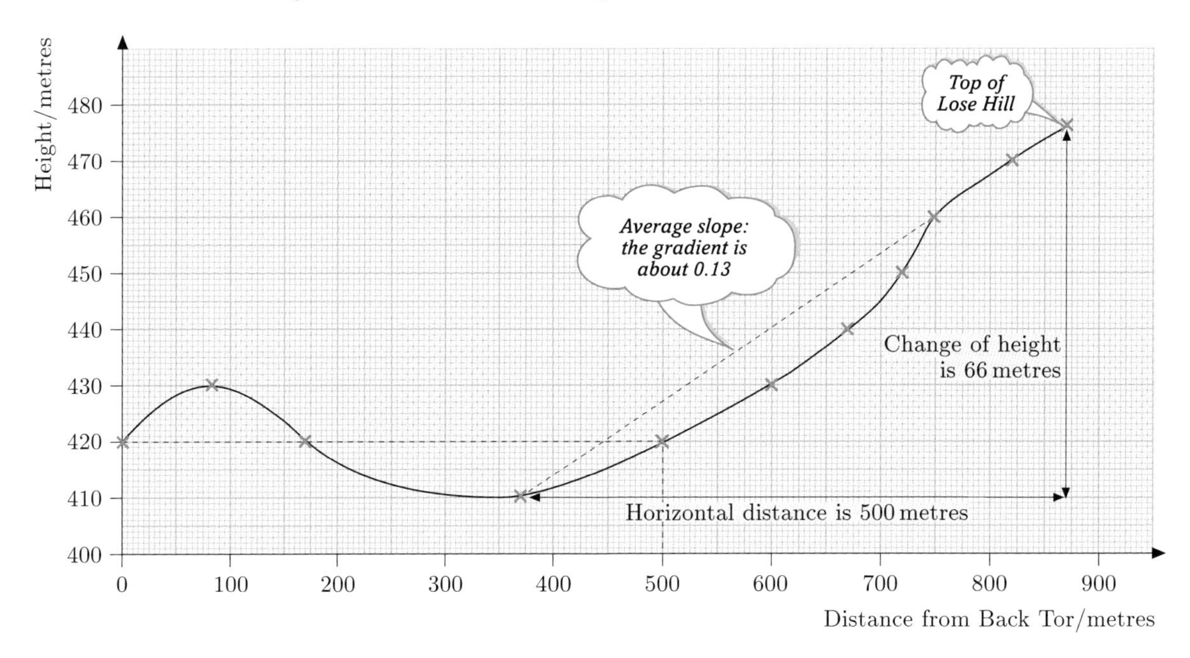

Figure 30 Completed graph of height plotted against distance

So average gradients can be misleading where the landscape is irregular.

Now return to the path earlier on the walk. From the top of Mam Tor, the ridge path descends to Hollins Cross. Table 4 gives the distance from Mam Tor and the corresponding height of the path.

Recall that the route of the walk follows the path along the ridge from Mam Tor to Hollins Cross.

Table 4 Distance and heights from Mam Tor to Hollins Cross

Distance from Mam Tor (in metres)	Height of path (in metres)
0	517
63	510
163	500
263	490
338	480
375	470
475	460
538	450
625	440
700	430
825	420
950	410
1100	400
1350	390

Activity 32 Plotting a profile

(a) Plot the information given in Table 4 on a graph, using centimetre graph paper. Plot *distance* on the horizontal axis and *height* on the vertical axis. You will need to decide on the scale of the axes, and label them appropriately.

(b) Give a detailed description of the path gradient from Mam Tor to Hollins Cross.

(c) Work out the average gradient for the stretch of path between Mam Tor and the boundary of the National Trust area, 800 m away.

You can now represent the entire walk from Mam Farm to Losehill Farm on a graph. Figure 31 shows a graph of height plotted against horizontal distance. The graph is drawn as if the path of the walk has been 'unfolded' and laid out with the start on the left and the finish on the right. You can see the relatively steep climbs up to Mam Tor and the shorter section at Back Tor.

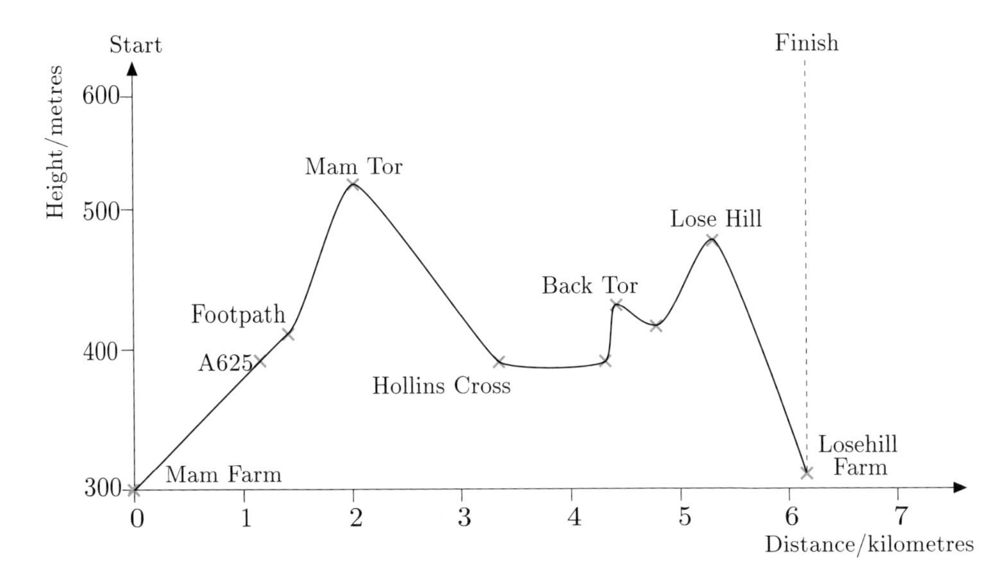

Figure 31 Plot of the profile of the walk

However, you may feel that the plot is not quite right because the slopes seem to be too steep compared to the walk you saw in the video. So what is wrong?

The visual impression of a graph depends on the scales used on the axes. Notice in this case that a change of height of 200 m is represented by the same length on the vertical axis as a distance of 2500 m on the horizontal axis. Indeed, if the horizontal axis were drawn to the scale used on the vertical axis, it would be twelve and a half times longer.

Every picture tells a story, and graphical 'pictures' are no exception. One story of this unit concerns decisions about what to display, and how to display it.

The effect of the different scale is to exaggerate the vertical profile of the walk, so that the slopes appear to be far steeper than they are in reality. Changing the vertical scale so that it is the same as the horizontal scale would reduce the distortion, but the trade-off is that variations in height become small and difficult to read from the graph. When you plot a graph, be aware of the visual effect of using different scales on the axes.

Activity 33 Seeing and believing

Look back at the graphs in this subsection. For each one, make a note about the effect of the scales of the axes on the visual impression given by the graph.

Make notes on what you need to consider when choosing the scales of axes.

4.3 Different slopes for different folks

So far you have been sketching graphs and working out gradients using the information on your OS map. This is fine if you are planning a walk, but gradient information occurs in other contexts. Road signs on steep hills usually give an indication of slope.

In the UK, the Department of Transport has used two different ways of showing the gradient of a road. Figure 32(a) shows an older sign: the gradient is given as a ratio '1 in 10'. Nowadays, gradient information is usually expressed as a percentage, as in Figure 32(b).

▶ What do these figures mean?

Figure 32 Road signs showing (left) a gradient of 1 in 10 and (right) a gradient of 20%

A slope of 1 in 10 means a change in height of 1 unit for every 10 units you travel. A slope of 20% means that the height of the road will change by 20 m for every 100 m travelled along it.

Activity 34 Road gradients

(a) What would be the change in height over a distance of 200 m if the gradient is 15%?

(b) What is the road gradient, expressed as a percentage, if the road increases in height by 50 m over 300 m?

Notice that there is a subtle difference between the definition of gradient in the previous subsection and the definition used by the road authorities.

Recall that the earlier definition of gradient was:

$$\text{gradient} = \frac{\text{increase in height}}{\text{horizontal distance}}.$$

This definition of gradient is shared by walkers and mathematicians and, in this section, it is called the *mathematical gradient*. Let's represent it by the symbol G_{M}. Then

$$G_{\mathrm{M}} = \frac{H}{D}$$

where H is the increase in height over a horizontal distance D.

In contrast the *road gradient*, on road signs, is calculated by dividing the increase in height of the road between two points by the distance measured *along the surface* of the road. So the definition is:

The road gradient is hence much easier to measure directly.

$$\text{road gradient} = \frac{\text{increase in height}}{\text{road surface distance}}.$$

So if the road surface distance is R, then the road gradient G_{R} is

$$G_{\mathrm{R}} = \frac{H}{R}.$$

But what relationship is there between the mathematical gradient and the road gradient? You can find out by using some mathematics.

Once again, use a model to help you focus on the main features. The model is very simple—in fact, it is just a triangle. You can think of the triangle in Figure 33 as the profile of a road going up a 'perfect' hill: there are no lumps or bumps and the road climbs with a constant gradient.

Recall the reader article in *Unit 1*: a cabbage is not a sphere, and a hill is not a triangle.

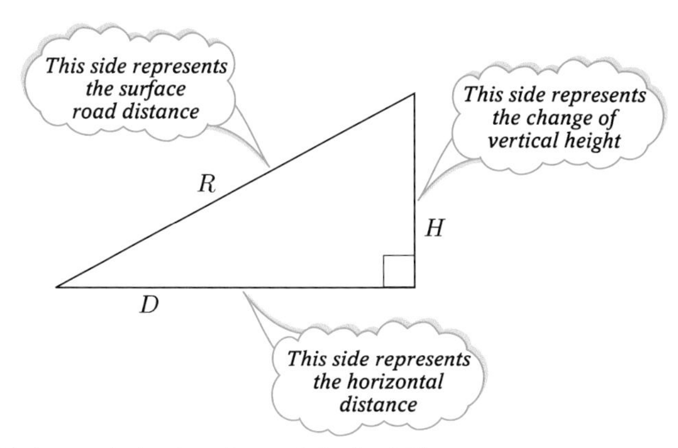

Figure 33 Right-angled triangle model of a hill

Notice that Figure 33 is a special sort of triangle. The angle between the horizontal side and the vertical side is 90 degrees or a *right angle*. Since the angles in any triangle always add up to 180 degrees, the other two angles in Figure 33 add up to 90 degrees (each must be less than 90 degrees). The right angle is opposite the longest side, which is called the *hypotenuse*.

The hypotenuse represents the road surface running up the hill. Its length is R. The horizontal side of the triangle represents the horizontal distance (found from a map) between two points on the road; its length is D. The vertical side represents the increase in height of the road; its length is H.

Information about the height increase and the horizontal distance can come from an OS map, but not the road surface distance. Finding the distance along the road corresponds in the mathematical model to finding the length of the hypotenuse in Figure 33. To do this you can use a well-known result in mathematics, called Pythagoras' theorem.

Pythagoras' theorem

The result called *Pythagoras' theorem* is one of the oldest mathematical results known, and one of the most famous. The theorem has long been attributed to Pythagoras, a Greek of the sixth century BC, who gave his name to a sect called the Pythagoreans. They followed a mystical philosophy, believing that numbers and number patterns were the key to understanding the world. However, it is clear from the clay tablets of about 2000 BC that early Babylonian scribes knew about 'Pythagoras' theorem', and the result is also found in ancient Chinese manuscripts.

What Greek mathematicians may have contributed was the idea of 'proving' that the theorem always holds. In this form, it appears in the first book of Euclid's Elements, a work written in Egypt around 300 BC, which drew together much earlier mathematical knowledge and which has remained one of the world's mathematical best-sellers ever since.

Pythagoras' theorem relates the lengths of the three sides of a right-angled triangle to one another. So if you know the lengths of any two sides, then you can always calculate the length of the third. It says:

> the square of the hypotenuse is equal to the sum of the squares of the other two sides.

For Figure 33 this means that:

$$(\text{road surface distance})^2 = (\text{horizontal distance})^2 + (\text{vertical height})^2$$

or, more concisely,

$$R^2 = D^2 + H^2.$$

Pythagoras' theorem crops up time and time again in mathematics. An important point to remember is that it applies *only* to right-angled triangles, like our model of a hill.

Here is a numerical example. On your map, locate the kilometre square 13 82 and within it the road from Castleton passing through Winnats. On the steepest section, the road rises by 50 m over a horizontal distance of about 300 m. So $H = 50$ m and $D = 300$ m and the mathematical gradient is

$$G_{\mathrm{M}} = \frac{H}{D} = \frac{50\ \mathrm{m}}{300\ \mathrm{m}} = 0.167\ (3\ \mathrm{d.p.}).$$

Now using Pythagoras' theorem you can work out the distance R along the surface of the road.

$$R^2 = D^2 + H^2 = (300\ \mathrm{m})^2 + (50\ \mathrm{m})^2 = 92\,500\ \mathrm{m}^2.$$

R is found by taking the square root of both sides.

$$R = \sqrt{92\,500\ \mathrm{m}^2} \simeq 304\ \mathrm{m}.$$

So the distance along the surface of the road is 4 metres greater than the horizontal distance. The road gradient is calculated as:

$$G_{\mathrm{R}} = \frac{H}{R} = \frac{50\ \mathrm{m}}{304\ \mathrm{m}} = 0.164\ (3\ \mathrm{d.p.}).$$

The road gradient (0.164) is very close to the mathematical gradient (0.167). For all practical purposes—and for all reasonable slopes that you are likely to come across on the road—the two are effectively the same. Here, both gradients indicate a slope of between 16 and 17 per cent.

However, the difference between the results of the calculations becomes greater as the slope gets steeper.

Example 7 How steep is Mam Tor?

At its steepest, the ground on the side of Mam Tor rises by 50 m over a horizontal distance of 25 m. Find the mathematical and road gradients.

The horizontal distance is $D = 25$ m and the vertical distance is $H = 50$ m. So the mathematical gradient is

$$G_{\mathrm{M}} = \frac{H}{D} = \frac{50\ \mathrm{m}}{25\ \mathrm{m}} = 2.$$

To find the corresponding road gradient, you need the distance R along the surface of the slope. Use Pythagoras' theorem.

$$R^2 = D^2 + H^2 = (25\ \mathrm{m})^2 + (50\ \mathrm{m})^2 = 3125\ \mathrm{m}^2.$$

Taking the square root gives: $R = \sqrt{3125\ \mathrm{m}^2} \simeq 56$ m. So the road gradient here would be:

$$G_{\mathrm{R}} = \frac{H}{R} = \frac{50\ \mathrm{m}}{56\ \mathrm{m}} \simeq 0.89 = 0.9\ (1\ \mathrm{d.p.}),$$

which is quite different from the value for the mathematical gradient of 2.

As the slope gets steeper the mathematical gradient gets bigger and bigger. The road gradient, however, gets closer and closer to 1.

Activity 35 Using Pythagoras' theorem

Suppose that from a map you have found that the horizontal distance between two points on a road going up a hill is 100 m, and that over this distance the height increases by 20 m.

(a) Use Pythagoras' theorem to find the distance measured along the road.

(b) Calculate the mathematical gradient and the road gradient of the road.

4.4 Estimating area

Various town planning Acts of Parliament in the 1920s and 1930s on land registration, land drainage, slum clearance and land valuation increased the demand for large-scale maps from which areas could be accurately determined.

Ordnance Survey maps can be used to estimate areas as well as distances.

A wide range of national and local government agencies, water authorities, land management and environmental bodies use area measurements made from maps. UK farmers may also use OS maps to calculate the areas of land they propose to use for particular crops. Others may need to work out areas from maps, perhaps to check development proposals. Finding an area from a map requires a little additional mathematics.

km^2 stands for square kilometres.

Blue grid lines divide your map into squares. Since the grid lines are spaced at 1 km intervals the area represented by one square is $1 \text{ km} \times 1 \text{ km}$, or 1 km^2.

▶ What is the total area covered by your map?

You could count all the grid squares, or more simply note that there are nine grid squares along the bottom and eight up the side, giving a total of $8 \times 9 = 72$ squares. So your map represents an area of 72 km^2.

You may find a local map or street plan with a scale is useful.

A square kilometre is quite a large area. Stop for a moment and think what would be included in 1 km^2 centred on where you are now.

For smaller areas, a square kilometre (km^2) is too large. A square metre (m^2) might be a more convenient measure.

▶ How many square metres are there in a square kilometre?

1 km is the same as 1000 m, so the area represented by one grid square is

$$1000 \text{ m} \times 1000 \text{ m} = 1\,000\,000 \text{ m}^2.$$

Ensure you multiply the units as well as the numbers.

So the area of one grid square is one million square metres or, in scientific notation, 10^6 m^2. So

$$1 \text{ km}^2 = 1\,000\,000 \text{ m}^2 = 10^6 \text{ m}^2.$$

Activity 36 Converting units of area

How many square centimetres (cm^2) are there in one square metre (m^2)?

▶ How are areas shown on your OS map related to areas on the ground?

By subdividing the kilometre grid squares on the map to produce smaller squares, you can estimate the areas of irregular-shaped features, such as woods and lakes, by counting the number of smaller squares inside it.

Each side of a kilometre grid square can be divided into ten equal parts. Drawing in the ten horizontal and ten vertical lines corresponding to these subdivisions, as in Figure 34, gives $10 \times 10 = 100$ smaller squares.

▶ What area does such a small square represent on the flat ground?

Each kilometre grid square represents an area of $1\ \text{km}^2$, or $10^6\ \text{m}^2$. So each of the 100 small squares will represent an area of $1/100\ \text{km}^2 = 0.01\ \text{km}^2$ or $10^6/10^2\ \text{m}^2 = 10^4\ \text{m}^2$.

Another way of looking at it is to note that each small square represents an area 100 m by 100 m, equal to $10\,000\ \text{m}^2$. This area is called a *hectare*, and there are 100 hectares in $1\ \text{km}^2$. The abbreviation for hectare is 'ha'.

You may like to add to your Handbook

$$1\ \text{km}^2 = 100\ \text{ha} = 10^6\ \text{m}^2.$$

A hectare is about 2.5 acres.

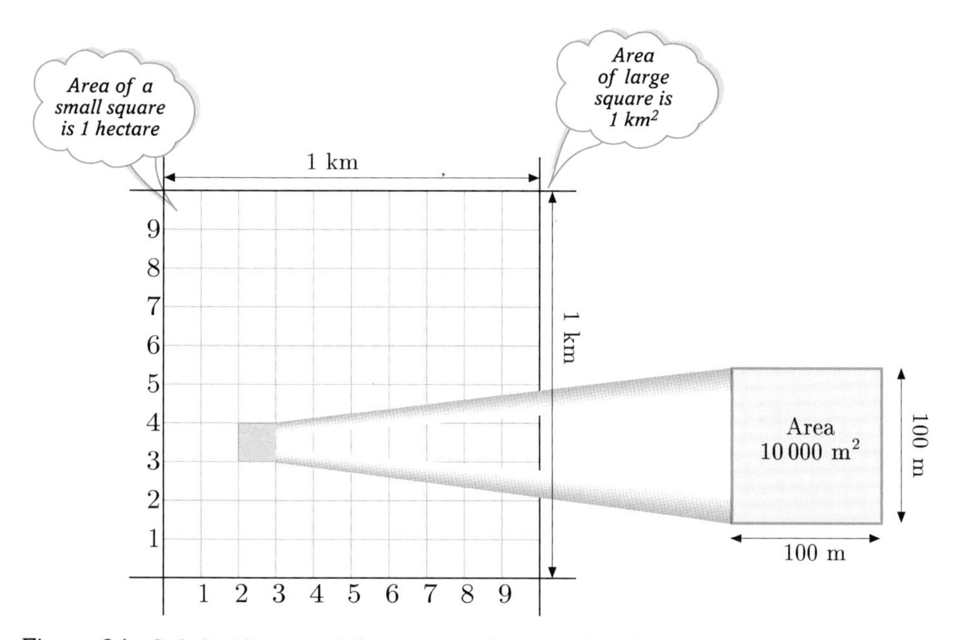

Figure 34 Subdividing a 1-kilometre grid square into hectares

Calculations of area made by subdividing grid squares and estimating the number of smaller squares covering a map feature will only be approximate.

Activity 37 Estimating area

Towards the top of your map in the grid square 14 89 is a small wood, called Gillott Hey Coppice. Estimate the area of this wood in hectares.

The scale of a map relates measurements of distance on the map to distances on the ground. For your 1 : 25 000 map, this relationship can be written as a formula:

distance on the ground = 25 000 × distance on the map

$$D_{\mathrm{G}} = 25000 \times D_{\mathrm{M}}.$$

The map legend does not tell us explicitly how areas are related. However, (for level ground) the area is proportional to the area on the map:

area on the ground = some number × area on the map.

▶ What number should go into this formula?

Map areas represent areas viewed directly from above. Like distances, the area of a sloping piece of ground will appear smaller the steeper the ground becomes. The area of a vertical cliff edge, to take an extreme example, would not have any area on the map at all.

The map scale shows that a distance of 1 unit *length* on the map represents 25 000 unit length on the ground. Now use this relationship to work out the *area* scaling factor. 1 square unit on the map represents 25 000 units × 25 000 units = 625 000 000 square units on the ground.

So a square 1 cm by 1 cm on the map represents a square 25 000 cm by 25 000 cm on the ground. In other words, 1 cm^2 on the map represents 25 000 cm × 25 000 cm = 625 000 000 cm^2 on the ground.

Thus if the map is drawn to a scale of 1 : 25 000 then the areas will be related by a scale of 1 : (25 000 × 25 000) or 1 : $(25\,000)^2$ or 1 : 625 000 000.

So the formula relating map and ground areas is:

area on the ground = 625 000 000 × area on the map

where both areas are measured in the same units. In symbols, this is

$$A_{\mathrm{G}} = (25\,000)^2 \times A_{\mathrm{M}} \quad \text{or} \quad A_{\mathrm{G}} = 625\,000\,000 \times A_{\mathrm{M}}$$

where A_{G} is the area on the ground and A_{M} is the area on the map, measured in the same units.

Activity 38 Predicting areas

(a) What area in hectares is represented by an area of 1 cm^2 on your 1 : 25 000 map?

(b) What area would 1 cm^2 represent if you were using a smaller scale 1 : 50 000 map?

Going from the map to the ground involves *multiplying* the map area by the area scaling factor. The other way round, you can relate ground areas to map areas by *dividing* the ground area by the area scaling factor.

On your OS map extract, the area scaling factor is $(25\,000)^2$. So

$$A_{\mathrm{M}} = A_{\mathrm{G}} \div (25000)^2 \quad \text{or} \quad A_{\mathrm{M}} = \frac{A_{\mathrm{G}}}{(25\,000)^2}.$$

So an area of 0.5 km^2 is represented on the map by an area of

$$A_\text{M} = \frac{0.5}{(25\,000)^2}\ \text{km}^2 = \frac{0.5}{(25\,000)^2} \times (1000)^2 \times (100)^2\ \text{cm}^2 = 8\ \text{cm}^2.$$

Activity 39 Representing areas

(a) The Ordnance Survey produces 1 : 2500 maps for detailed planning. What area on the map represents an area of 1 ha on the ground?

(b) If the map scale were 1 : 1250, what map area would represent 1 ha?

This section has shown how height and distance information from a map can be used to plot a profile of the ground to show the slopes.

Subsection 4.1 looked at gradient as a mathematical description of slope. Subsection 4.3 discussed two definitions of gradient: one for map-users and one for road authorities. Mathematicians and map-users adopt the same definition of gradient: the ratio of increased vertical height to horizontal distance (calculated directly from height and distance information on an OS map). The road gradient calculation needs the road surface distance. Pythagoras' theorem was used to calculate this distance from the height and distance information given by the map.

Subsection 4.4 used grid squares to predict area. Each large square bordered by blue grid lines represents an area of 1 km^2. For smaller areas, divide the grid squares into 100 smaller squares. Each represents an area of 10 000 m^2, or 1 ha. The conversion scale between map area and ground area is given by the square of the map scale.

Before starting Section 5, check you have all the notes for your Handbook.

Outcomes

After studying this section, you should be able to:

- ◇ calculate the average gradient of a slope from a map (Activities 29 and 30);
- ◇ draw and interpret a profile using map data on height and position (Activities 31–33);
- ◇ use the following terms accurately and be able to explain them to someone else: 'mathematical gradient', 'road gradient', 'positive gradient', 'negative gradient', 'Pythagoras' theorem', 'hypotenuse', 'profile', map 'area scale' (Activities 29, 30, 32–36 and 38);
- ◇ use Pythagoras' theorem to work out the length of the hypotenuse of a right-angled triangle and hence predict road gradients from a map (Activity 35);
- ◇ convert between area units and relate an area measured on a map to an area measured on the ground (Activities 36–39).

5 A calculated display

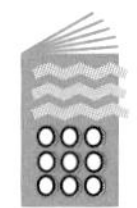

Aims The main aim of this section is to show how your calculator can be used for plotting line graphs, and for performing calculations using lists of coordinate data. This section also aims to consolidate your map work. ◇

A list of numbers may be the starting point for a plot of the profile of a hillside, or for repeated calculations such as scaling map measurements or estimating times using Naismith's rule.

Plotting graphs point-by-point or repeating the same sort of calculation over and over again is tedious and time consuming when there are a lot of data. But help is at hand. Once you have lists of numbers in your calculator, you can display line graphs and carry out repeated calculations quickly and with a lot less effort. But the calculator does not do all the work for you! You must still enter the data correctly, and choose the range over which you want the graph to be displayed. The calculator does not label the axes, so you must also remember what the display represents.

Now work through Chapter 6 of the Calculator Book.

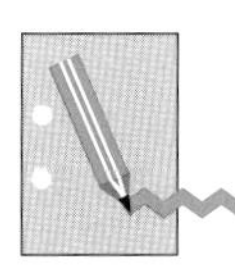

Activity 40 Reviewing your progress

Now that you have completed this unit, think about your own progress so far.

(a) First, look at your activity and Handbook entries. Are you surprised by the range of mathematical techniques and activities you have covered as you worked through the unit? A number of key terms have been introduced in the unit. Take time to check you have notes on the definitions for your Handbook.

(b) You have used a number of media to study this unit, including written text with activities, an OS map, reader articles, video, audio and the calculator. You may also have attended a tutorial session, worked with other students, or telephoned your tutor.

Think about all the media you have used as part of your learning in this unit. For each one, try to say *how* it helped you to learn the mathematics and to come to grips with the ideas.

From your study of the course so far have you found some media components to be more effective than others in helping you to learn? Why do you think they have been more effective?

How did your planning for your study of this unit work out?

(c) Look back at your answer to Activity 1. These skills and understanding will also be needed in the next few units. Consider what you could now add concerning:

(i) your skills with the course calculator

(ii) your understanding of the concept of ratio.

Outcomes

You should now be able to use your calculator to:

◇ display a line graph in an appropriate window, given a data list of coordinate pairs;

◇ perform calculations on the numbers in one or more data lists and display the results as a list.

You should be developing skills in:

◇ using 'review activities' as part of learning (Activity 40).

Unit summary and outcomes

This unit has been about representations and relationships related to maps. To convey meaning, a representation must be read and interpreted, and to do this you need to be able to make sense of the symbols and conventions used. Mathematics itself can be viewed as a language of special representations—numbers, symbols, diagrams—used to express particular ideas and relationships.

Section 1 looked at some of the different ways maps are made to represent the world. A map will stress some aspects and ignore others. Map representations involve a rich structure of mathematical relationships as well as cartographic tradition and practice.

Section 2 brought out some of the mathematical ideas embedded in an OS map. These included a two-dimensional coordinate system for specifying position, scales to convert between distances on the map and on the ground, and contour lines representing heights.

Section 3 focused on mathematical models and formulas. They were used to represent relationships, predict distances, compass bearings and the time that a walk will take.

Section 4 continued the modelling theme and discussed the representation of area on a map and different definitions of gradient. It introduced Pythagoras' theorem to relate the lengths of the sides of a right-angled triangle. You also saw how height and distance information can be presented graphically—and used graph-drawing techniques to provide a visual impression of the profile of the ground.

In Section 5 you used your calculator, bringing modelling and graphical skills together and consolidating your learning. In planning the walk in Section 3, you completed a route card with grid references, distance and height information for each stage of the walk. Using Naismith's rule you were able to work out the time for each part of the walk separately, and also predict the time for the whole walk. This involved the repeated use of Naismith's rule. Using the facilities built into your calculator for doing arithmetic on lists of numbers, the repeated calculations can be done together.

The unit has pointed out to you that a map, like *any* representation, stresses some features but ignores others. Keep this in mind when you meet other symbolic graphic and diagrammatic representations in the rest of the course. Learning mathematics successfully involves handling symbolic, graphical and numerical conventions.

Outcomes

You should now be able to:

◇ review your learning and knowledge of a topic to help plan your studies;
◇ read maps, networks, articles and unit text for a purpose and extract appropriate information;
◇ describe the properties that cannot all be preserved in a transformation from a 3D to a 2D representation;
◇ identify features in a representation or other mathematical model that have been stressed and those that have been ignored;
◇ use grid references to specify a location on an OS map;
◇ convert measurements of distances made on the ground to their corresponding values on a map and vice versa;
◇ explain how contour line patterns represent three-dimensional topography on a two-dimensional map;
◇ use a protractor to measure grid bearings on an OS map;
◇ predict a compass bearing given a grid bearing and information about magnetic north, and vice versa;
◇ fix a location using the bearings from two other points;
◇ use a formula such as that for Naismith's rule;
◇ calculate the average (mathematical) gradient of a slope from a map;
◇ use Pythagoras' theorem to calculate the length of the hypotenuse of a right-angled triangle and hence a road gradient;
◇ draw and interpret a profile from map data on height and position;
◇ relate an area measured on a map to an area measured on the ground;
◇ display on your calculator a line graph in an appropriate window given a list of coordinate pairs;
◇ perform calculations on the numbers in one or more lists entered in your calculator, and display the results as a list;
◇ use the following terms accurately and be able to describe their meaning and use to others: 'compass bearing', 'grid bearing', 'grid north', 'grid system', 'map scale', 'contour line', 'key', 'legend', 'magnetic north', 'mathematical model', 'mathematical gradient', 'road gradient', 'positive gradient', 'negative gradient', 'hypotenuse', 'profile' and 'map area scales'.

Appendix: Using a protractor to measure angles

This appendix helps you to use a protractor, in conjunction with listening to band 1 of CD5509.

Frame 1

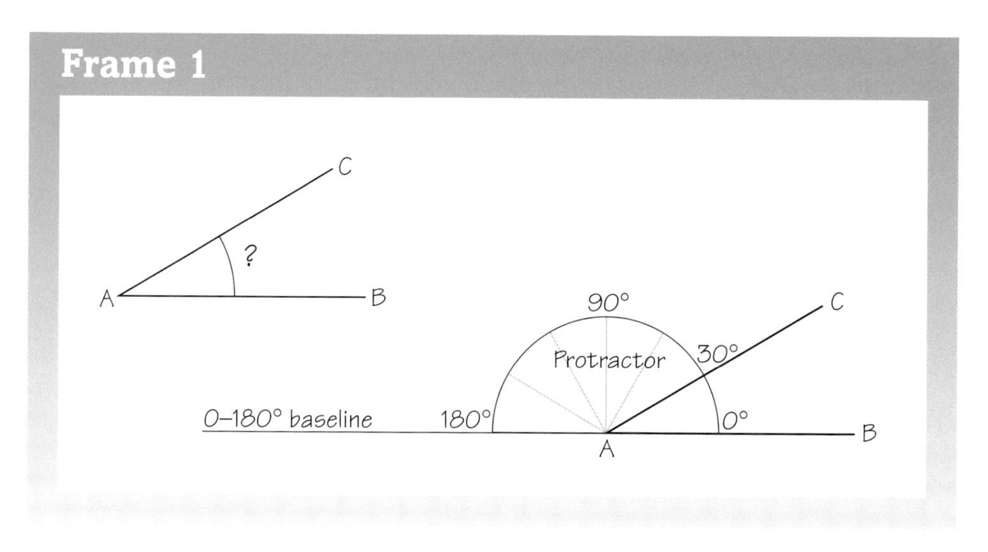

Frame 2

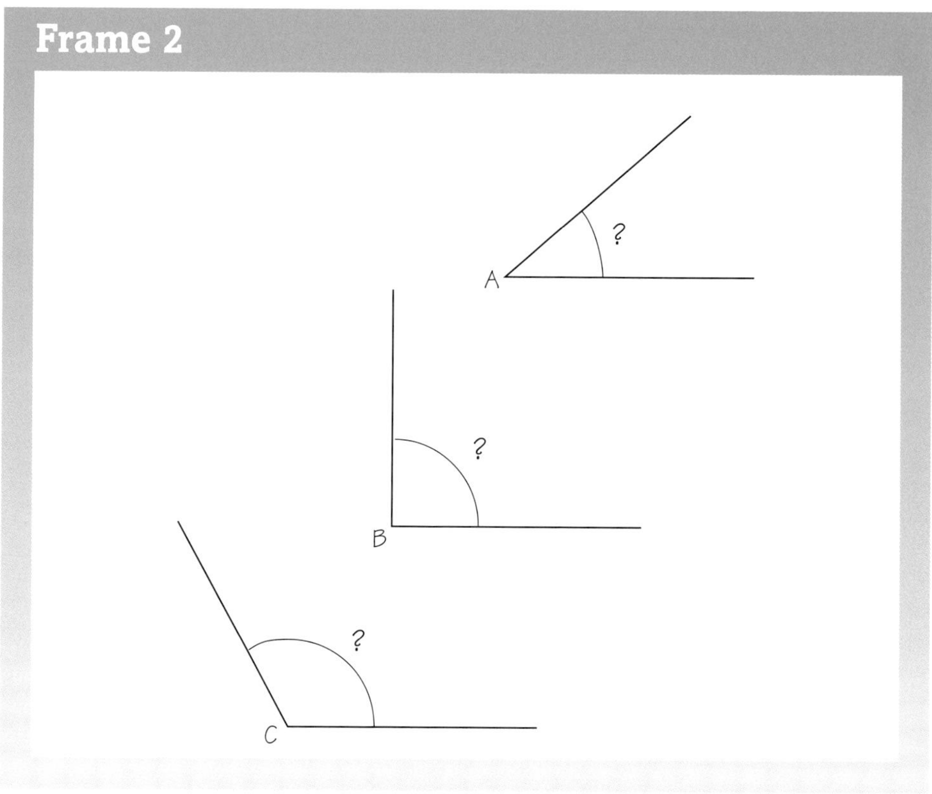

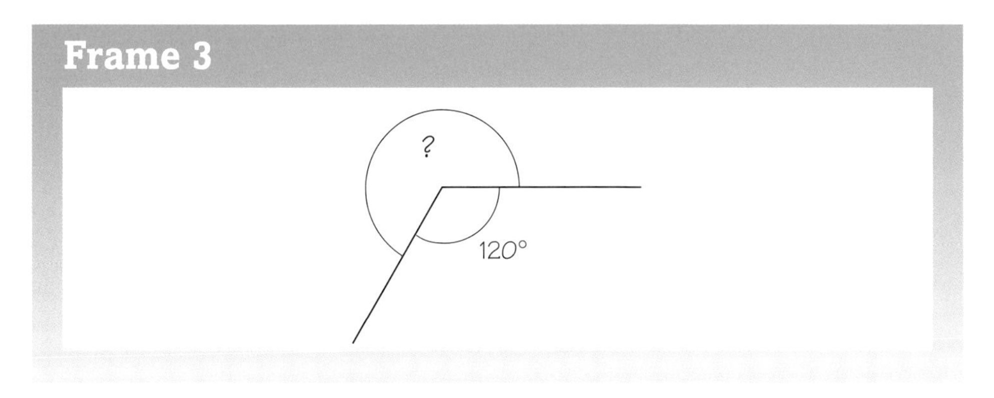
Frame 3
?
120°

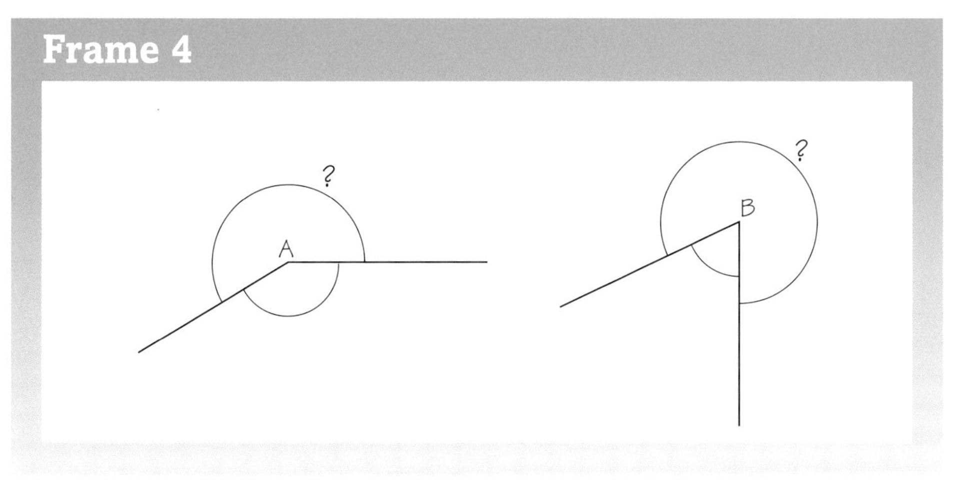
Frame 4
?
A
?
B

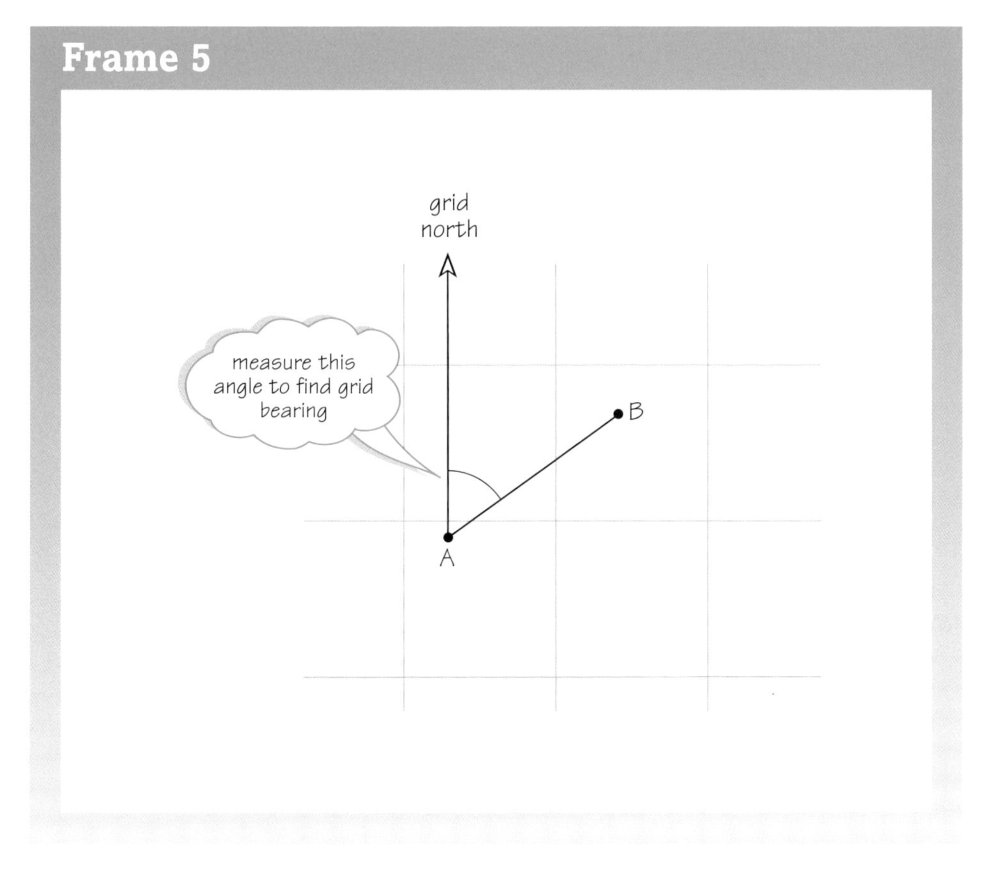
Frame 5
grid
north
measure this
angle to find grid
bearing
B
A

Frame 6

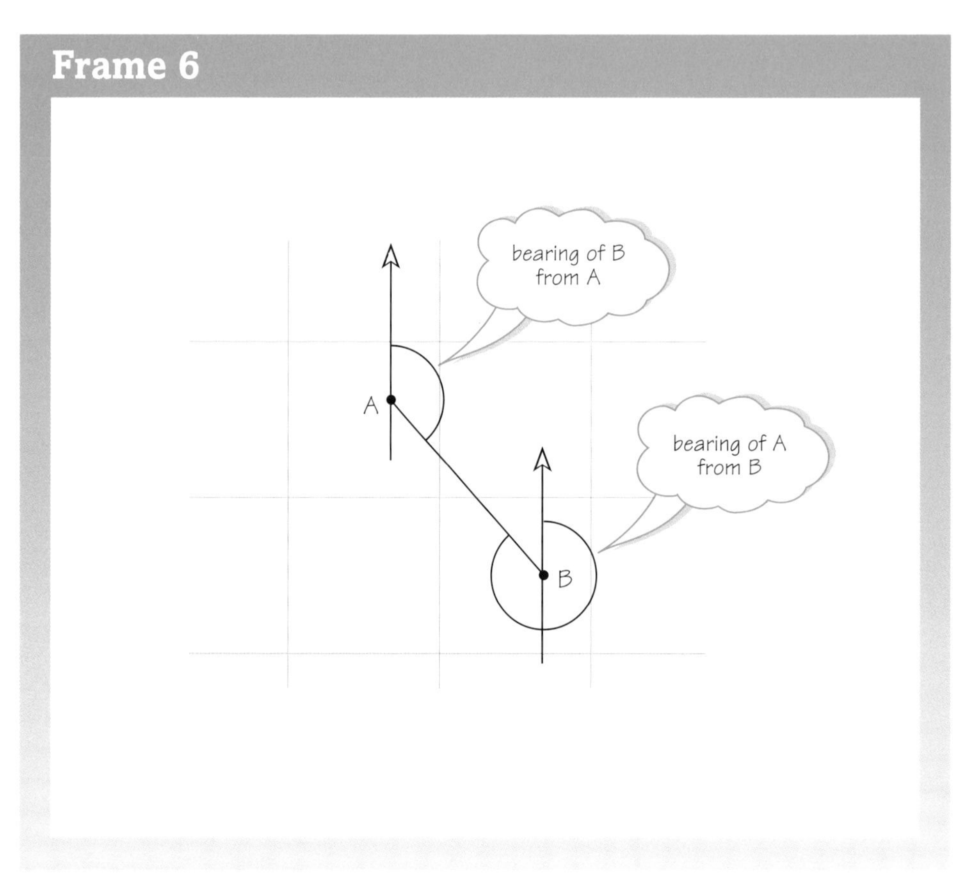

Frame 7

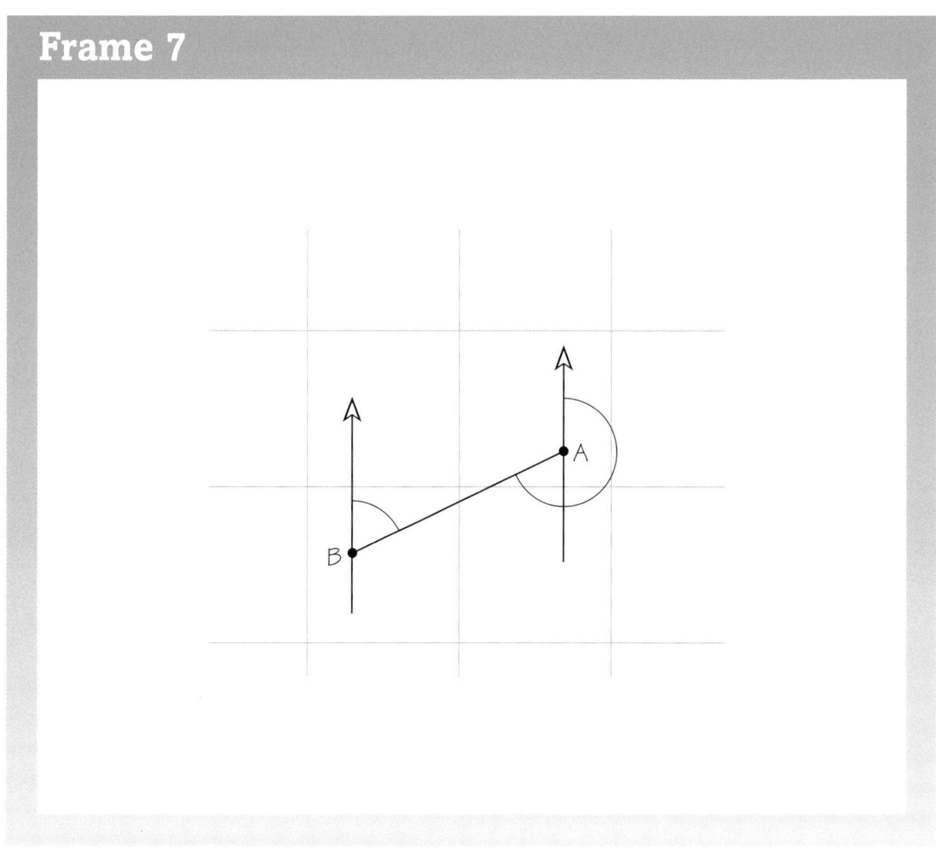

Comments on Activities

Activity 1

(a) Everybody's answer will be different, but you should have addressed the following points.

(i) Your statistical skills before studying Block A, perhaps calculating the mean; your graphical skills before studying Block A, perhaps drawing graphs by hand; your calculator skills before studying Block A, perhaps arithmetic calculation.

(ii) The skill and the level you have acquired during Block A, with reference to where you learnt them, for example inputting data into lists and then using STAT CALC to find mean, median, quartiles, range; using STAT PLOT to plot a frequency chart, boxplots, scatter plots. (Give relevant references from the Calculator Book Chapters 2, 3, 4 and 5 TMAs and/or CMAs.)

(iii) You might mention the notes you take, in order to recall key sequences for doing these things; comments from your tutor; tutorial activities; or brain stretchers. Any of these factors may have helped you.

Also mention any factors that hindered your progress with the calculator.

(b) (i) You may have chosen examples of your own work from any of the previous units, perhaps a TMA or CMA answer, a unit activity or an exercise from the calculator book or resource book. They should show how a ratio is calculated and how it is used.

(ii) A ratio is one quantity divided by another to give a relative comparison. You might talk about percentage changes, rates, scale diagrams (Unit 0) proportion and scaling, price ratios (Unit 2), earning ratios (Unit 3), relative spread (Unit 4). You may have discussed why the use of ratios in relative comparisons gives a fairer comparison than an absolute comparison.

Activity 2

Everybody's answers will be different—there is no one single answer. Map (a) shows the road systems around the Open University campus in Milton Keynes. Someone could use this map to find a route when visiting the OU.

You may initially notice that map (b) is an extract from a published Ordnance Survey map. Historially such maps were made for military purposes. However, if you enjoy walking, you might use this type of map to plan a walk and then navigate on the walk.

Map (c) seems to have religious symbols and map (d) might have been designed with a political purpose. The orientation of maps (c) and (d) may have surprised you. You might use them to think about how the world looks from different points of view.

Map (e) is the sort of sketch map which many people draw to help somebody follow a route to a particular place. It contains what they consider to be important landmarks.

Activity 3

The bearings are: 45 degrees for north-east, 135 degrees for south-east, 225 degrees for south-west, and 315 degrees for north-west.

Activity 4

When designing and producing a map, there are always decisions to be made about which elements of the 3D situation you wish to preserve in 2D. These elements can include the following.

- Shape: the shapes of coast lines, and so on, are preserved.
- Distance: preserving distance means that the whole map is drawn to the same scale so that equal distances on the map correspond to equal distances on the ground.

◇ Area: preserving area means that the whole map is drawn so that equal areas on the map correspond to equal areas on the ground.

◇ Orientation: consistent orientation means that the whole map is drawn so that any particular orientations on the map always correspond to the same orientations on the ground.

◇ Direction: preserving direction means that straight-line directions on the map (the 2D image) correspond to straight lines on the actual ground.

You may have thought of other examples of transformations from 3D to 2D, such as photographs, television, cinema, scans and X-ray images.

Activity 5

(a) Your answer to this will depend upon where you live, but you would probably try to draw angles roughly correctly and to include important features of her journey from the bus stop. Scale may vary, and detail too. There might be no need to have north at the top. However, you would need some indication of orientation, so that she starts off in the right direction from the bus stop.

(b) Preserving directions would be important and, as far as possible, drawing the map to scale would be desirable. The North Pole would need to be in the middle of the map, so north cannot be at the top. An alternative method of orientating the pilot would be needed. The lines of latitude and longitude might be useful.

(c) North is conventionally at the top. However, distances on the map do not reflect changes in height of the ground, the third dimension here. Contour lines, used to indicate heights, are discussed later in the unit.

(d) There may be a conflict here between preserving direction, distances and shapes in mapping a sphere onto a flat map. The choices you make will depend upon your own cultural background and priorities. You could mark the journeys on one of the world projections discussed in this section, or you could use a different one.

Activity 6

(a) B to C is 4 km, C to E is 9 km. So B to E is 13 km.

(b) B to A is 5 km, A to G is 7 km, G to F is 3 km and F to E is 1 km. The total is $(5 + 7 + 3 + 1)$ km. So B to E is 16 km.

Activity 7

A map that is designed to be read by finger needs relatively poor resolution compared with one designed to be read by eye. Thus, a radical change of approach is needed by the map designers to include only relevant information about the route.

Activity 8

World maps of the occurrences of different modes of transmission of the AIDS virus are used by researchers into the origins and spread of the disease.

Lack of detail may be partly explained by concern with confidentiality. The subdivisions in Britain are too large to enable useful predictions to be made for health planning.

Activity 9

Wood's article reinforces some of the messages in this section. He comments that anyone who fails to see that a map is 'a weapon disguised as an impartial survey of the way things are' falls victim to it. Images of the world, however accurate, are bound to stress certain features and ignore others. And when you interpret maps, or indeed any image, you need to be aware of what data were collected, who the map was created for, the purpose it was designed for, and so on.

Activity 10

If you have not used an Ordnance Survey map before, you may need to work through this section quite slowly, perhaps coming back a second time to some topics to make sure you understand them. However, if you are already familiar with OS maps, then the section will give you a chance to practise your skills, with an eye for the mathematical aspects of map reading.

Look quickly through the section, noting which topics will be new to you. How does this affect your study plan?

Activity 11

Using Figure 14, the grid squares are:

(a) NN (200 700)

(b) TG (600 300)

The grid references are:

(c) 300 000 (SY)

(d) 500 200 (TL)

Activity 12

The grid reference is made up of 127, the easting, and 836, the northing.

The '12' of the easting and the '83' of the northing locate the south-western corner (bottom left) of a 1 km square. Locate the grid square that has the easting of 12 on the left, and the northing of 83 running along the bottom.

The given location square lies inside this square: the smaller square is defined by the digit 7 of the easting and the digit 6 of the northing.

By eye, or using a pencil and ruler, divide each of the sides of the 1 km grid square into 10 divisions. Mentally draw seven divisions from the left along the bottom of the square and six divisions from the bottom up the left-hand edge. The lines cross at the south-western corner of the grid reference square. In that small square you should find a small triangular symbol, indicating the triangulation pillar at the top of Mam Tor.

Activity 13

The grid references are as follows.

(a) Hollins Cross 136845

(b) Back Tor (path crossing) 144848

(c) Lose Hill 153853

Activity 14

The map references correspond to the following.

(a) 172834 church

(b) 181832 railway station

(c) 123832 picnic site

(d) 140866 youth hostel

(e) 187851 spot height 462 m at Winhill Pike

Activity 15

(a) Distance on the map is $D_M = 12$ cm. So actual distance on the ground is:

$$D_G = 25\,000 \times 12 \text{ cm} = 300\,000 \text{ cm}$$
$$= \frac{300\,000}{100} \text{ m} = 3000 \text{ m} = 3 \text{ km}.$$

(b) Distance on the ground is $D_G = 7.5$ km which is 7 500 m. So the map distance is:

$$D_M = \frac{7500}{25\,000} \text{ m} = 0.3 \text{ m} = 30 \text{ cm}.$$

Activity 16

You should have found a map distance of about $D_M = 4.5$ cm.

This represents a ground distance $D_G = 25\,000 \times 4.5 \text{ cm} = 112\,500 \text{ cm} = 1125$ m, or 1.125 km.

Alternatively, you could use the fact that 4 cm on the map represents 1 km on the ground. So a map distance of 4.5 cm represents a ground distance of 4.5/4 km = 1.125 km.

You might round this distance to 1.1 km.

Activity 17

Hollins Cross to Back Tor, on the map, is $D_{\mathrm{M}} = 4.3$ cm, corresponding, on the ground, to

$$\begin{aligned} D_{\mathrm{G}} &= 25\,000 \times 4.3 \text{ cm} = 107\,500 \text{ cm} \\ &= 1.075 \text{ km} = 1.1 \text{ km (1 d.p.)}. \end{aligned}$$

Back Tor to Lose Hill, on the map, is $D_{\mathrm{M}} = 3.5$ cm, corresponding, on the ground, to

$$\begin{aligned} D_{\mathrm{G}} &= 25\,000 \times 3.5 \text{ cm} = 0.875 \text{ km} \\ &= 0.9 \text{ km (1 d.p.)}. \end{aligned}$$

Lose Hill to Losehill Farm, on the map, is $D_{\mathrm{M}} = 3.4$ cm, corresponding, on the ground, to

$$\begin{aligned} D_{\mathrm{G}} &= 25\,000 \times 3.4 \text{ cm} = 85\,000 \text{ cm} = 0.85 \text{ km} \\ &= 0.9 \text{ km (1 d.p.)}. \end{aligned}$$

The total distance from Mam Farm to Losehill Farm on the map is about $D_{\mathrm{M}} = 24.5$ cm, corresponding, on the ground, to

$$\begin{aligned} D_{\mathrm{G}} &= 25\,000 \times 24.5 \text{ cm} = 612\,500 \text{ cm} \\ &= 6.1 \text{ km (1 d.p.)}. \end{aligned}$$

Alternatively, if you add up the distances in kilometres in Table 1, you will get 6.2 km. (The difference is due to rounding errors). So the total distance is about 6.1 or 6.2 km.

Activity 18

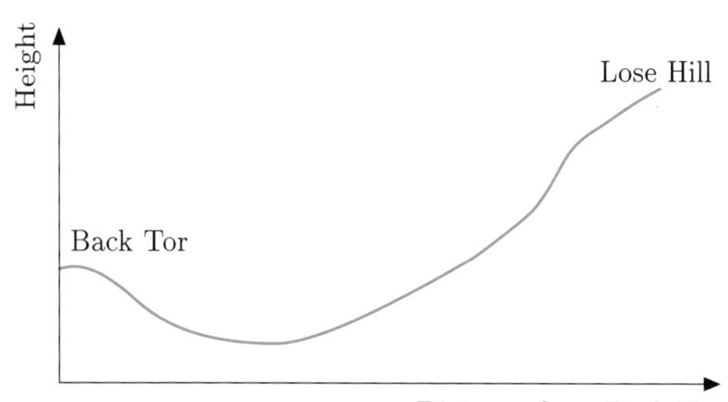

The figure shows a sketch of the path from Back Tor to Lose Hill. Midway between Back Tor and Lose Hill the contours are relatively widely spaced and around the 417-metre spot height the path is level. Approaching Lose Hill the line of the path cuts the contour lines, which get closer together, indicating that the steepness of the path increases towards the summit. For a sketch profile, do not be too concerned with including every detail relating to height. A sketch is, as the word implies, a visual impression.

Activity 19

Everybody will have something slightly different, but here are some ideas.

(a) You could use the compass extracts from the video together with a protractor, a compass and a map.

You should include the following.

- ◇ Horizontal angles and directions are preserved on the map, but magnetic north needs to be taken into account for accurate bearings.
- ◇ Align the compass magnetic needle with the magnetic north on the map.

(b) You could use the video extracts of Mam Tor and Hollins Cross with the computer animation and description of contour lines. You could liken a saddle point to a horse's saddle and have pictures of both. Include the following.

- ◇ Heights are represented by contour lines on the map; when the lines are close together the slope is steep.
- ◇ It is important to notice directions in which contour heights are increasing and decreasing in order to identify features like peaks or dips and which way the 'saddle' goes.

Activity 20

(a) $Y = 18$. So $M = 5 - \frac{18}{6} = 5 - 3 = 2$. So the formula predicts that magnetic north will be 2° west of grid north in 2012.

(b) Every 6 years the difference decreases by 1 degree. So another 12 years are needed after 2012, i.e. 2024. This is 30 years after 1994. Check in the formula with $Y = 30$,

$$M = 5 - \frac{30}{6} = 5 - 5 = 0.$$

Activity 21

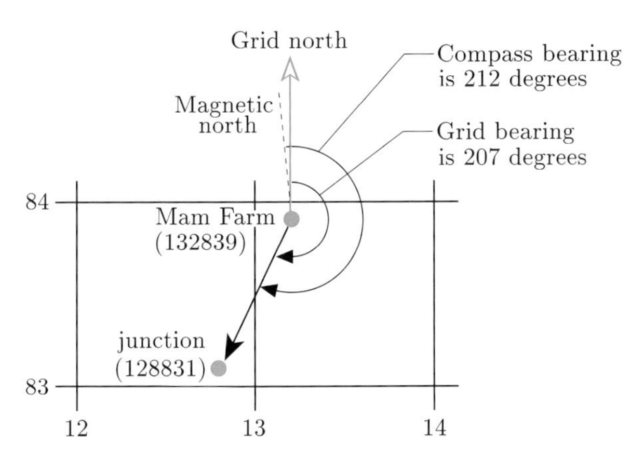

From the map, the junction lies on a grid bearing of about 207°. Going from map to ground means adding 5° to the grid bearing, so the compass bearing should be 212°.

Activity 22

Measuring the angles from grid north gives the bearing of Upper House Farm as about 26°, and of Blackden View Farm as about 67°. Since you are going from the map to the ground, add 5° to each. This predicts that the compass bearings, *when measured in the physical world* (in 1994), will be 31° and 72° respectively.

Activity 23

From Blackden View Farm the compass bearing will be $72° + 180° = 252°$.

Activity 24

The reverse compass bearings are $18° + 180° = 198°$ from Upper House Farm, and $61° + 180° = 241°$ from Blackden View Farm. These are converted to grid bearings by *removing* 5° from each, giving a grid bearing of 193° from Upper House Farm and a grid bearing of 236° from Blackden View Farm.

If you draw the lines corresponding to these bearings on your map, as shown in the figure at the top of the next column, you should find that they cross at about 116885, indicating that you are a little further south-east than you had planned.

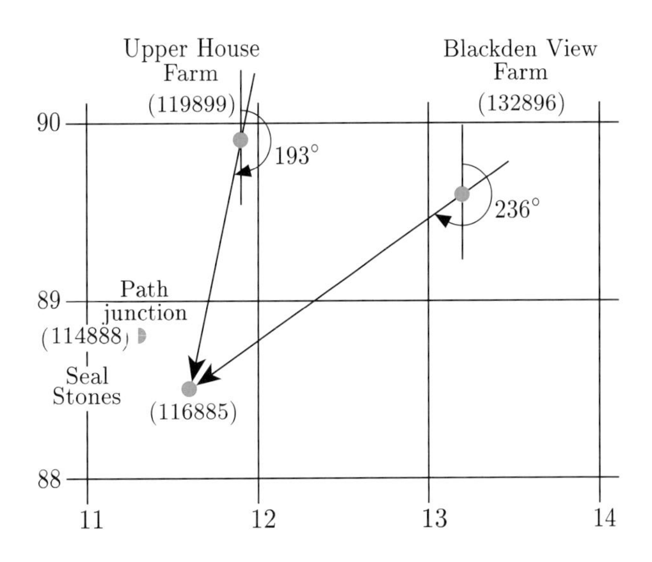

Activity 25

The grid bearing is about 145°. Adding 5° to account for magnetic north predicts a compass bearing of some 150° (in 1994).

Activity 26

The walk is 2 km long. So $D = 2$.

The ground rises by $517\text{m} - 300\text{m} = 217\,\text{m}$. So $H = 217$. Naismith's rule predicts:

$$T = \frac{2}{5} + \frac{217}{600} = 0.4 + 0.36 = 0.76.$$

Now 0.76 hours is about 46 minutes, but most walkers would round this up and estimate 50 minutes for the time taken to complete the walk.

Activity 27

In your list you may have identified that Naismith's rule:

◇ stresses the same constant speed for walking across level ground and going down slopes;

◇ stresses a slower constant speed when walking uphill.

But it also ignores:

◇ the fact that people get tired—it assumes that walking and climbing speeds will be the same throughout the walk;

- variations in what people carry;
- weather conditions;
- ground conditions;
- differences in fitness and stamina between groups of walkers.

In your Handbook notes do include the rule

$$T = \frac{D}{5} + \frac{H}{600}$$

stating clearly what T, D and H stand for and the units in which they must be measured.

Activity 28

From Back Tor to Lose Hill, the total climb is about 70 m, over a distance of about 0.9 km. So $D = 0.9$ and $H = 70$. Naismith's rule predicts

$$T = \frac{D}{5} + \frac{H}{600} = \frac{0.9}{5} + \frac{70}{600} \simeq 0.18 + 0.12 = 0.30.$$

This in hours; in minutes it is 0.30×60, or about 20 minutes.

The path down from Lose Hill to Losehill Farm is again about 0.9 km long. So $D = 0.9$. Naismith's rule ignores the height descended.

So $H = 0$ and $T = \dfrac{0.9}{5} + 0 = 0.18$ (hours).

This is $0.18 \times 60 = 11$ minutes (to the nearest minute). You might want to round up to 15 minutes.

Adding the times for the stages gives $(50 + 20 + 20 + 20 + 15)$ minutes $= 125$ minutes, or just over 2 hours for the complete walk.

From the start of the walk at Mam Farm to the finish at Losehill Farm the distance is approximately $D = 6.2$ (km). The total ascent is about $H = 327$ (m). Naismith's rule predicts that the time to complete the walk, in hours, will be

$$T = \frac{D}{5} + \frac{H}{600} = \frac{6.2}{5} + \frac{327}{600} \simeq 1.8.$$

This is about 1.8 hours, but adding on time for stops to admire the view, catch your breath, drink some water and so on, supports the previous estimate of about 2 hours.

Activity 29

(a) Measuring in metres, the gradient is $100 \text{ m}/300 \text{ m} = 0.33$ (to 2 d.p.). The ground rises 0.33 m for every metre moved horizontally.

(b) Measuring in cm, the change of height is $100 \times 100 \text{ cm} = 10^4$ cm and the horizontal distance is $300 \times 100 \text{ cm} = 3 \times 10^4$ cm.

So the gradient is $\dfrac{10^4 \text{ cm}}{3 \times 10^4 \text{ cm}} = 0.33$ (2 d.p.).

(c) Measuring in km, the change of height is $\dfrac{100}{1000}$ km $= 0.1$ km, and the horizontal distance is $300/1000$ km $= 0.3$ km.

So the gradient is $\dfrac{0.1 \text{ km}}{0.3 \text{ km}} = 0.33$ (2 d.p.).

(d) Multiplying (or dividing) the top and bottom of a fraction or ratio by the same number does not change the value of the ratio.

Activity 30

Over a distance of $D = 200$ m the height of the ground falls by 150 m. The height is decreasing rather than increasing with distance. So $H = {}^{-}150$ m in this case and the calculation is:

$$G = \frac{H}{D} = \frac{{}^{-}150 \text{ m}}{200 \text{ m}} = {}^{-}0.75.$$

The negative gradient is interpreted to mean that, on this side of Mam Tor, the ground falls by just over 0.75 m (75 cm) for every metre of horizontal distance.

Activity 31

The completed graph is in Figure 30.

Zero on the horizontal scale represents the position of Back Tor. The height of the path increases to 430 m just beyond Back Tor, before dipping to 410 m about 300 m further on. The graph then slopes up to the top of Lose Hill. The graph is not a straight line, however, so the gradient is not constant. The path is steepest about 700 m from Back Tor.

Activity 32

(a)

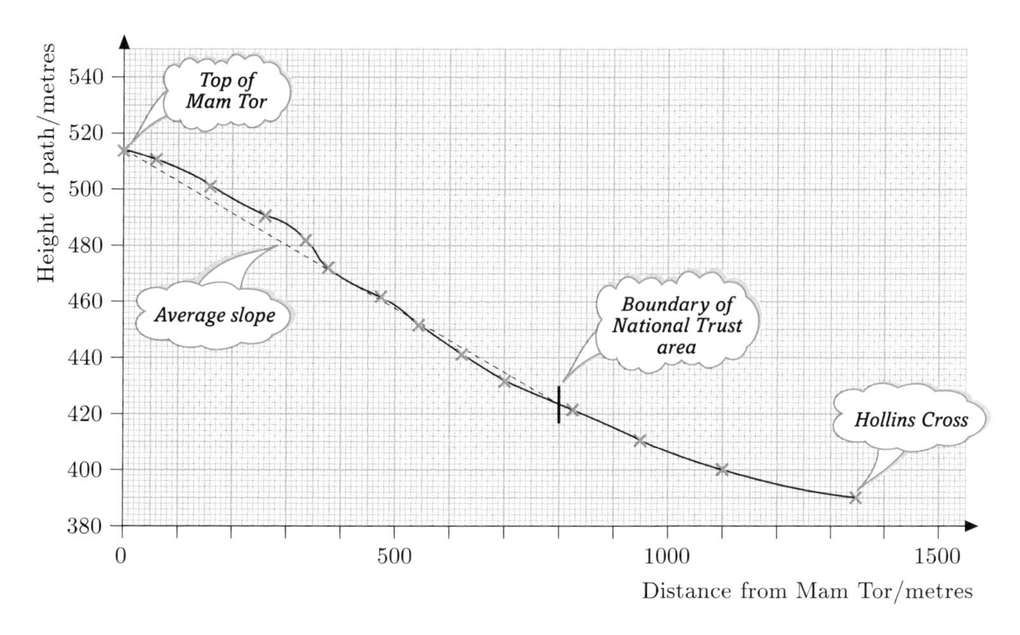

Figure 35 Profile of the path from Mam Tor to Hollins Cross

(b) The path slopes down from Mam Tor with a gentle steady gradient for about 340 m, then it steepens slightly for a short distance, before resuming the original slope. It starts to flatten out approaching Hollins Cross.

(c) Between Mam Tor and the National Trust boundary the path drops from 517 m to about 425 m, over a distance of 800 m. The average gradient of this stretch will be negative (because the height decreases) and is about
$(425 - 517) \div 800 = {}^{-}92/800 \simeq {}^{-}0.12$.

Activity 33

Figures 29 and 30 somewhat exaggerate the steepness, as does Figure 35 above, because the vertical and horizontal scales are not the same.

When choosing scales for your axes, consider:

- fitting the graph on the paper comfortably;
- ease of plotting and reading points;
- the effects of using different scales on the two axes.

Activity 34

(a) The change in height will be 15% of 200 m, which is $\frac{15}{100} \times 200\text{ m} = 30\text{ m}$.

(b) Expressing 50 as a percentage of 300 gives
$\frac{50}{300} \times 100\% \simeq 17\%$.

Activity 35

(a) By Pythagoras' theorem

$$R^2 = D^2 + H^2.$$

$D = 100$ m and $H = 20$ m, so

$$R^2 = (100\text{ m})^2 + (20\text{ m})^2$$
$$= 10\,000\text{ m}^2 + 400\text{ m}^2$$
$$= 10\,400\text{ m}^2.$$

R is found by taking the square root:

$$R = \sqrt{10\,400\text{ m}^2}$$
$$\simeq 102\text{ m}.$$

(b) The mathematical gradient is:

$$G_M = \frac{H}{D} = \frac{20}{100} = 0.2$$

The road gradient is:

$$G_R = \frac{H}{R} = \frac{20}{102} \simeq 0.196.$$

Either way, the road steepness is about 20%.

Activity 36

Imagine a square 1 m by 1 m. A metre contains 100 cm, so in one square metre there are:

$100 \text{ cm} \times 100 \text{ cm} = 10\,000 \text{ cm}^2 = 10^4 \text{ cm}^2$.

A square metre is ten thousand square centimetres.

Activity 37

On the map, between five and six small squares (each representing $1 \text{ ha} = 10\,000 \text{ m}^2$) can be fitted into the coppice area. So the area is about 5 to 6 ha.

Activity 38

(a) 1 cm^2 represents an area on the ground of $1 \times 25\,000^2 \text{ cm}^2 = 625\,000\,000 \text{ cm}^2$. There are $10\,000 \text{ cm}^2$ in 1 m^2, so the ground area is $625\,000\,000/10\,000 \text{ m}^2 = 62\,500 \text{ m}^2$.

$1 \text{ ha} = 10\,000 \text{ m}^2$. So the ground area $= 62\,500/10\,000 \text{ ha} = 6.25 \text{ ha}$.

(b) 1 cm^2 represents an area on the ground of $1 \times 50\,000^2 \text{ cm}^2 = 2\,500\,000\,000 \text{ cm}^2$. This is 25 ha (or 0.25 km^2).

Notice that *changing* the map scale ratio by a factor of two from 1 : 25 000 to 1 : 50 000 leads to a four-fold increase in ground area for the same map area because $2^2 = 4$.

Activity 39

(a) For a map with a scale of 1 : 2500, the area scaling factor is $1 : (2500)^2$, or 1 : 6 250 000. $A_G = 1 \text{ ha} = 10\,000 \text{ m}^2$.

So the area on the map is $A_M = 10\,000/6\,250\,000 \text{ m}^2 = 0.0016 \text{ m}^2$, or 16 cm^2.

(b) The map scale is 1 : 1250. 1 ha is $100 \text{ m} \times 100 \text{ m} = 10000 \text{ m}^2 = A_G$. So

$$A_M = \frac{A_G}{(1250)^2} = \frac{10000 \text{ m}^2}{1250 \times 1250} = \frac{10000 \times 10000 \text{ cm}^2}{1250 \times 1250} = 64 \text{ cm}^2.$$

The same ground area is represented by a map area four times larger than on the map with half the scale.

Activity 40

(a) You should have notes on most of the outcomes listed on page 67.

(b) Different people find different media components more or less useful. If you find some media components difficult to learn from, then consider what you could do to remedy this. For example, you might allow extra time for such components in planning your study of future units.

(c) (i) You should now be able to do arithmetic with data lists (storing the results in lists) and plot line graphs from data lists.

(ii) You should now be able to relate the concept of ratio to map scales (distance and area) and to gradients.